The Design of Public Places

城市空间与景观设计 4

[美] 约翰·莫里斯·狄克逊　编著

中国建筑工业出版社

著作权合同登记图字：01-2006-4601 号

图书在版编目（CIP）数据

城市空间与景观设计 4/（美）狄克逊编著；程素荣译. —北京：中国建筑工业出版社，2006
ISBN 978-7-112-08765-5

Ⅰ.城… Ⅱ.①狄… ②程… Ⅲ.①城市空间－建筑设计 ②城市－景观－设计 Ⅳ.TU984.11

中国版本图书馆 CIP 数据核字（2006）第 103506 号

Copyright © 2006 by Visual Reference Publications, Inc.
All rights reserved. No part of this book may be reproduced in any form or by any electronic or mechanical means, including information storage and retrieval system, without permission in writing from the publisher.

本书由美国 VRP 出版公司授权出版

责任编辑：程素荣

城市空间与景观设计 4
[美] 约翰·莫里斯·狄克逊 编著
*
中国建筑工业出版社出版、发行（北京西郊百万庄）
新 华 书 店 经 销
北京嘉泰利德公司制版
北京盛通彩色印刷有限公司印刷
*
开本：965×1270 毫米 1/16 印张：19¾ 字数：600 千字
2007 年 3 月第一版 2007 年 3 月第一次印刷
定价：266.00 元
ISBN 978-7-112-08765-5
(15429)

版权所有 翻印必究
如有印装质量问题，可寄本社退换
（邮政编码 100037）
本社网址：http://www.cabp.com.cn
网上书店：http://www.china-building.com.cn

Contents

Contents by Project Type	4
Introduction by John Morris Dixon, FAIA	7
Preface by Richard M. Rosan, FAIA, President, ULI	8
180° Design Studio	9
Altoon + Porter Architects LLP	17
annex|5, An Epstein Design Group	25
Austin Veum Robbins Partners	33
Beeler Guest Owens Architects, L.P.	41
Callison Architecture, Inc.	49
Canin Associates	57
Carter & Burgess, Inc.	65
Chan Krieger & Associates	73
Charlan • Brock & Associates, Inc.	81
CBT/Childs Bertman Tseckares Inc.	89
Costas Kondylis and Partners LLP	97
Cunningham + Quill Architects PLLC	105
Dahlin Group Architecture Planning	113
David M. Schwarz	121
Dougherty Schroeder & Associates, Inc.	129
Duany Plater-Zyberk & Company	137
Elkus-Manfredi Architects	145
ELS	153
Field Paoli	161
FXFOWLE ARCHITECTS, PC	169
Glatting Jackson Kercher Anglin Lopez Rinehart, Inc.	177
Goody Clancy	185
Lessard Group Inc.	193
Looney Ricks Kiss	201
MBH Architects	209
McLarand Vasquez Emsiek & Partners	217
Pappageorge/Haymes Ltd.	225
Perkowitz + Ruth Atchitects	233
Retzsch Lanao Caycedo Architects	241
RTKL	249
Sasaki Associates, Inc.	257
SEH (Short Elliott Hendrickson Inc.)	265
SWA Group	273
Swaback Partners pllc	281
Thomas Balsley Associates	289
Thomas P. Cox: Architects, Inc.	297
Project Credits	305
Index by Project	314
Acknowledgments	316

Contents by Project Type

Note: Most projects in this book include, by their nature, more than one function. A few guidelines to categories below:

Mixed-Use Developments: reserved for those with substantial mix of uses within buildings (beyond accessory parking or retail included in buildings listed as office, residential, etc.).
Communities: wide mix of uses, typically on cleared land, with new buildings and infrastructure.
Urban Redevelopment: large in scope, including existing construction and infrastructure.
Remodeling/Re-use: including specifics on re-use of individual buildings.
Other project types: largely devoted to listed type (not listing, for instance, inclusion of recreation in residential projects, plazas with office buildings, or street improvements in urban redevelopment).

Communities/Neighborhoods

Alys Beach, Panama City, FL, *144*
Amelia Park, Fernandina Beach, FL, *139*
Ave Maria Town Center, Collier County, FL, *206*
Biosphere 2, Oracle, AZ, *284*
Botanica on the Green, Stapleton, CO, *298*
Brighton Village, Coolidge, AZ, *32*
Buena Vista, Bakersfield, CA, *60*
Coyote Valley, San Jose, CA, *114*
Crescent Creek, Raytown, MO, *14*
DC Ranch, Scottsdale, AZ, *282*
Douglas Park, Long Beach, CA, *223*
Fortnightly Neighborhood Master Plan, New Senior Community Center, Herndon, VA, *112*
Habersham, Beaufort, SC, *141*
Heart of Anoka Commuter Rail Village Master Plan, Anoka, MN, *268*
Huntfield, Charles Town, WV, *107*
I'On, Mount Pleasant, SC, *140*
Luxe Hills International Golf Community, Cheng Du, Sichuan Province, China, *116*
Marana Master Plan, Marana, AZ, *288*
Marina University Villages, Marina (Fort Ord), CA, *212*
National Harbor, National Harbor, MD, *200*
New Longview, Lee's Summit, MO, *12*
Parker Square, Flower Mound, TX, *128*
Planning Design of Hailou West Coast, Haikou City, China, *28*
Promenade Town Center, Pasco County, FL, *64*
Rosemary Beach, Panama City, FL, *142*
Ross Bridge Village Center, Birmingham, AL, *207*
Solivita, Poinciana, FL, *62*
Southlake Town Square, Southlake, TX, *126*
Springfield Town Center, Springfield, VA, *196*
Tannin, Orange Beach, AL, *138*
Thu Thiem New Urban Center, Ho Chi Minh City, Vietnam, *260*
University Villages, Marina, CA, *118*
Victoria Gardens, Rancho Cucamonga, CA, *166*
Village of Kohler, The, Kohler, WI, *286*

Cultural Facilities

California Theatre, San Jose, CA, *154*
Fort Worth Master Plan, Fort Worth, TX, *124*
Lincoln Center Redevelopment, New York, NY, *173*

Educational Facilities

National Cathedral School, Washington, DC, *110*
Ottawa University Master Plan, Ottawa, KS, *15*
Whitman School of Management, Syracuse University, Syracuse, NY, *172*

Entertainment Complexes

Fashion Show, Las Vegas, NV, *22*
FedExForum, Memphis, TN, *204*
LaQua Tokyo Dome City, Tokyo, Japan, *250*
North Allston Strategic Framework for Planning, Boston, MA, *192*
On Broadway, Downtown Redwood City, CA, *164*
Power Plant Live!, Baltimore, MD, *69*
Walk, The, Atlantic City, NJ, *68*

Government Complex

Mound Public Safety Facility, Mound, MN, *266*

Hotel/Resort/Conference Centers

Dosflota Multipurpose Complex Master Plan, Moscow, Russia, *175*
Grande Lakes Resorts, Orlando, FL, *58*
Las Palomas, Puerto Penasco, Mexico, *287*
Rarity Pointe Lodge and Spa, Knoxville, TN, *86*
Tianjin Tower, Tianjin, China, *174*

Mixed-Use Developments

100 Cambridge Street/Bowdoin Place, Boston, MA, *150*
Allegro Tower Apartments, San Diego, CA, *34*
Ayala Center Greenbelt, Makati City, Philippines, *50*
Beverly Canon Mixed-Use Retail, Beverly Hills, CA, *162*
Caton's Walk, Georgetown, Washington, DC, *108*
Church Street Plaza, Evanston, IL, *156*
Dosflota Multipurpose Complex Master Plan, Moscow, Russia, *175*
Egyptian Lofts, San Diego, CA, *36*
Fifth Avenue Place, Phase II, Boca Raton, FL, *246*
Figueroa Central, Los Angeles, CA, *236*
Fruitvale Village, Oakland, CA, *218*
Gateway Centre, Longmont, CO, *272*
Glen Town Center, The, Glenview, IL, *226*
Grand Gateway, Shanghai, China, *52*
Hollywood & Vine, Hollywood, CA, *221*
LaQua Tokyo Dome City, Tokyo, Japan, *250*

North Hills, Raleigh, NC, **66**
On Broadway, Downtown Redwood City, CA, **164**
Park Laurel on the Prado, San Diego, CA, **38**
Pinnacle Museum Tower, The, San Diego, CA, **40**
Promenade at Rio Vista, The, San Diego, CA, **220**
Residences at Kendall Square, The, Cambridge, MA, **92**
Smart Corner, San Diego, CA, **39**
Thornton Park, Orlando, FL, **208**
Town Center at Levis Commons, The, Perrysburg (Toledo), OH, **214**
Uptown Maitland West, Maitland, FL, **85**
Victoria Gardens, Rancho Cucamonga, CA, **20**
Village of Merrick Park, The, Coral Gables, FL, **158**

Office Buildings / Developments
100 Cambridge Street/Bowdoin Place, Boston, MA, **150**
35 and 40 Landsdowne Street, University Park at MIT, Cambridge, MA, **146**
Cypress Park West, Phase II, Fort Lauderdale, FL, **244**
Destin Commons, Destin, FL, **130**
Forum at Sunnyvale, The, Sunnyvale, CA, **134**
Lite-On Electronic Headquarters, Taipei, Taiwan, **274**
Motorola Global Software Group, Krakow, Poland, **26**
New York Times Building, The, New York, NY, **171**
Pinnacle Hills Promenade, Rogers, AK, **132**
Royal Palm Office Building, Boca Raton, FL, **242**
Scottsdale Hangar One, Scottsdale, AZ, **285**
Serta International Center, Hoffman Estates, IL, **30**
Tianjin Tower, Tianjin, China, **174**

Parks
Addison Circle Park, Addison, TX, **258**
Broad Street Park, Baldwin Park, FL, **180**
Charleston Waterfront Park, Charleston, SC, **262**
Detroit Riverfront Civic Center Promenade, Detroit, MI, **264**
Hangzhou HuBin Commerce & Tourism District Redevelopment Master Plan, Hangzhou, China, **278**
Hollis Garden, Lakeland, FL, **178**
Parker Square, Flower Mound, TX, **128**
Riverside Park South, New York, NY, **295**

Plazas/Squares
100 Cambridge Street/Bowdoin Place, Boston, MA, **150**
35 and 40 Landsdowne Street, University Park at MIT, Cambridge, MA, **146**
Capitol Plaza, New York, NY, **292**
City Hall Plaza Community Arcade and Government Center Master Plan, Boston, MA, **76**
Detroit Riverfront Civic Center Promenade, Detroit, MI, **264**
J-City, Tokyo, Japan, **290**
Lakes at Thousand Oaks, The, Thousand Oaks, CA, **240**
Lewis Avenue Corridor, Las Vegas, NV, **279**

Lite-On Electronic Headquarters, Taipei, Taiwan, **274**
Pacific Design Center, West Hollywood, CA, **294**
PPG Place, Pittsburgh, PA, **276**
World Trade Center Plaza, Osaka, Japan, **296**

Remodeling / Re-use
100 Cambridge Street/Bowdoin Place, Boston, MA, **150**
Biosphere 2, Oracle, AZ, **284**
California Theatre, San Jose, CA, **154**
Caton's Walk, Georgetown, Washington, DC, **108**
Davis Building, The, Dallas, TX, **47**
Glen Town Center, The, Glenview, IL, **226**
Lincoln Center Redevelopment, New York, NY, **173**
Mather Building, The, Washington, DC, **106**
Power Plant Live!, Baltimore, MD, **69**
Village of Kohler, The, Kohler, WI, **286**

Residential Developments
100 Cambridge Street/Bowdoin Place, Boston, MA, **150**
200 Brannan, San Francisco, CA, **216**
4025 Connecticut Avenue, Park Hill Condominiums, Washington, DC, **111**
5225 Maple Avenue, Dallas, TX, **44**
600 North Lake Shore Drive, Chicago, IL, **228**
Allegro Tower Apartments, San Diego, CA, **34**
ALTA, The, Thomas Circle, Washington, DC, **109**
Aqua Condominiums, Panama City Beach, FL, **84**
Black Diamond, Pittsburg, CA, **120**
Block X, 1145 Washington, Chicago, IL, **229**
Botanica on the Green, Stapleton, CO, **298**
Cheval Apartments on Old Katy Road, Houston, TX, **88**
Crescent Park Apartment Homes, Playa Vista, CA, **300**
Davis Building, The, Dallas, TX, **47**
Easton, Dallas, TX, **46**
Echelon I, Las Vegas, NV, **237**
Egyptian Lofts, San Diego, CA, **36**
Fifth Avenue Place, Phase II, Boca Raton, FL, **246**
Flats at Rosemary Beach, The, Rosemary Beach, FL, **87**
Grand Tier, The, 1930 Broadway, New York, NY, **104**
Gulf Coast Town Center, Fort Myers, FL, **136**
Helena Apartment Building, The, New York, NY, **170**
Heritage, The, New York, NY, **100**
Highlands of Lombard, The, Lombard, IL, **42**
Jefferson at Providence Place, Providence, RI, **202**
Kinzie Park, Chicago, IL, **232**
Las Palomas, Puerto Penasco, Mexico, **287**
Mather Building, The, Washington, DC, **106**
Metropolitan Tower, Seattle, WA, **56**
Milan Lofts, Pasadena, CA, **237**
Morton Square, 600 Washington St., New York, NY, **102**

Park Laurel on the Prado, San Diego, CA, **38**
Pinnacle Museum Tower, The, San Diego, CA, **40**
Pointe at Middle River, The, Oakland Park, FL, **248**
Rarity Pointe Lodge and Spa, Knoxville, TN, **86**
Residences at Kendall Square, The, Cambridge, MA, **92**
Residential Projects, **16**
Rollins Square, Boston, MA, **93**
Smart Corner, San Diego, CA, **39**
Tianjin Tower, Tianjin, China, **174**
Trump World Tower, 845 United Nations Plaza, New York, NY, **98**
Union Station, Union, NJ, **48**
Uptown Maitland West, Maitland, FL, **85**
Williams Walk at Bartram Park, Jacksonville, FL, **82**

Shopping Centers
Avenue East Cobb, The, Atlanta, GA, **133**
Ayala Center Greenbelt, Makati City, Philippines, **50**
Beverly Canon Mixed-Use Retail, Beverly Hills, CA, **162**
Bridgeport Village, Tualatin, OR, **234**
Buena Park Downtown, Buena Park, CA, **239**
Citrus Plaza, Redlands, CA, **70**
Destin Commons, Destin, FL, **130**
Fashion Show, Las Vegas, NV, **22**
Grand Gateway, Shanghai, China, **52**
Gulf Coast Town Center, Fort Myers, FL, **136**
Knox Shopping Centre, Melbourne, Australia, **18**
Lakes at Thousand Oaks, The, Thousand Oaks, CA, **240**
Mercantile West, Ladera Ranch, CA, **238**
North Hills, Raleigh, NC, **66**
On Broadway, Downtown Redwood City, CA , **164**
Pinnacle Hills Promenade, Rogers, AK, **132**
Principe Pio, Madrid, Spain, **254**
Scottsdale Hangar One, Scottsdale, AZ, **285**
Shoppes at Blackstone Valley, The, Millbury, MA, **71**
Streets of Tanasbourne, The, Hillsboro, OR, **168**
Suwon Gateway Plaza, Suwon, Korea, **54**
Town Center at Levis Commons, The, Perrysburg (Toledo), OH, **214**
Victoria Gardens, Rancho Cucamonga, CA, **20**
Walk, The Atlantic City, NJ, **68**
Walkers Brook Crossing, Jordan's Furniture, Home Depot, IMAX, North Reading, MA, **72**
West Hollywood Gateway, West Hollywood, CA, **210**

Sports / Recreational Facilities
FedExForum, Memphis, TN, **204**
Luxe Hills International Golf Community, Cheng Du, Sichuan Province, China, **116**

Streetscape Improvements
I-35WS Access Project, Hennepin County, **271**
Lewis Avenue Corridor, Las Vegas, NV, **279**
Loring Bikeway and Park, City of Minneapolis, MN, **270**
Park Avenue Streetscape, Winter Park, FL, **182**
Santana Row, San Jose, CA, **280**

Transportation/Highway Facilities
Fort Washington Way Highway Reconfiguration, Cincinnati, OH, **80**
Fruitvale Village, Oakland, CA, **218**
I-35WS Access Project, Hennepin County, **271**
Loring Bikeway and Park, City of Minneapolis, MN, **270**
Promenade at Rio Vista, The, San Diego, CA, **220**
Scottsdale Hangar One, Scottsdale, AZ, **285**
Suwon Gateway Plaza, Suwon, Korea, **54**
West Village, Dallas, TX, **127**

Urban Redevelopment
Assembly Square, Somerville, MA, **190**
Black Diamond, Pittsburg, CA, **120**
Canton Crossing, Baltimore, MD, **198**
Church Street Plaza, Evanston, IL, **156**
City of Sunnyvale Downtown Design Plan, The, Sunnyvale, CA, **160**
Columbus Center, Boston, MA, **90**
Douglas Park, Long Beach, CA, **223**
Downtown Brea Development District, Brea, CA, **252**
Fort Point Channel, Boston, MA, **186**
Fort Worth Master Plan, Fort Worth, TX, **122**
Forum at Sunnyvale, The, Sunnyvale, CA, **134**
Grand Avenue Competition, Los Angeles, CA, **302**
Hangzhou HuBin Commerce & Tourism District Redevelopment Master Plan, Hangzhou, China, **278**
Kinzie Park, Chicago, IL, **232**
Marana Master Plan, Marana, AZ, **288**
Museum Park, Chicago, IL, **230**
New Town Theater District, St. Charles, MO, **10**
North Allston Strategic Framework for Planning, Boston, MA, **192**
North Point, Cambridge, Boston, Somerville, MA, **91**
Parker Square, Flower Mound, TX, **128**
Prudential Center Redevelopment, The, Boston, MA, **94**
Renaissance Place Redevelopment Plan, Naugatuck, CT, **176**
The Heights, Tampa, FL, **184**
Three Rivers Park, Pittsburgh, PA, **78**
Tralee, Dublin, CA, **224**
Trump Plaza, New Rochelle, NY, **194**
Uptown Oakland Development, Oakland, CA, **222**
Village of Kohler, The, Kohler, WI, **286**
Village of Merrick Park, The, Coral Gables, FL, **158**
West Village, Dallas, TX, **127**
Westgate – Pasadena, Pasadena, CA, **304**
Zha Bei/The Hub International Lifestyle Centre, Shanghai, China, **256**

Introduction

The key characteristics of good urban development can be distilled from a review of current projects.

In their various ways, the projects presented in this book revitalize the public realm that has been sapped for decades by dispersed development, dominance of the automobile, and misguided planning and zoning policies. As I've examined all of these 150-plus projects, certain of the most promising qualities of urban development for this new century have come to the fore.

One widely accepted goal of urban projects is to achieve a mix of functions that will generate round-the-clock activity. Attendant benefits include the opportunity to walk, not drive, from home to work or shopping and the reduction in total resources devoted to parking – with many of the spaces serving residents at night and workers and shoppers by day. Mixed use also establishes a day-and-evening economic basis for a richer variety of retail and dining facilities. There are rarely good restaurants, at any price point, in districts occupied only by day workers.

Mixing pedestrians and vehicles on the same right of way, reversing decades of planning dogma, has now become a criterion for good urban development – or even good shopping center remodeling. Few people, it turns out, enjoy vast parking lots or garages, but we must not forget that vehicles do little to enhance the pedestrian experience, so the setting shared by pedestrians and vehicles demands subtle design. And some pedestrian-only precincts, which are found in admired historical cities, are still desirable.

A well-designed urban complex projects a strong sense of place. It has memorable architectural elements and/or spatial compositions of its own, and it effectively incorporates – at best capitalizes on – local attributes of climate, topography, and traditions.

Many projects are able to benefit from the reuse of existing structures. The best incorporate them without papering over their design peculiarities, their accumulated scars, or the dissonant juxtapositions that often come with adaptation to new circumstance. No theme-park restorations, please.

Good community or urban redevelopment plans make effective connections with the surrounding community, including but not limited to knitting their streets into existing networks. The kinds of uses accommodated should complement the surrounding area, without undermining existing activities and without creating jarring social shifts at project boundaries. Admittedly, real-world pressures often make such ideal relationships hard to achieve.

Where a project is one element of a larger revitalization process, there is no question that it must be designed to support that larger set of goals, at best exceeding the requirements of the plans they contribute to.

Good urban projects support public transportation. Density and mix of uses are keys to aggregating enough riders to support a bus or rail line. Clustering higher-density residential around stops or stations also supports adjoining commerce, which in some notable cases makes the transportation node a destination in itself.

Promotion of economic and social integration is an objective of many of our finest urban developments. The separation of people by economic strata, while not totally avoidable, can be reduced in many creative ways. Many of the new communities and neighborhoods presented in this book encourage an economic mix by offering a wide variety of residential units within a small area – studio apartments, lofts, "granny" units over garages – in a fine-grained mix with more lavish residences. It has been proven over and over that, given a sensitively designed environment, affluent residents will happily opt to live among people with a wide range of resources and lifestyles.

Sustainability has become something of a buzz word lately, but there is no question that it is the obligation of all designers – all people – to conserve resources and reduce waste. While one of the proven ways is to encourage walking instead of driving – or even taking the train – there are many other architectural and planning strategies. Much of the savings of energy and material resources are the province of mechanical engineers and product designers, but architects can do much to reduce the demand for artificial lighting, air conditioning, and heating, while choosing building components wisely. Planners and landscape architects can make valuable contributions to minimizing traffic and pollution, promoting natural ventilation, directing rainwater runoff, and providing for natural shade.

Somewhat sadly, many of our most creative urban efforts involve mitigating the effects of earlier public work by adapting thoroughfares built to please traffic engineers to make them appealing to pedestrians, by making freeways more compatible with surrounding neighborhoods, by dividing monumental vacant plazas into places of human scale and activity, and by carving attractive open spaces out of massive building clusters. As the following pages indicate, we are mending our past urban ways at the same time we are creating new environments that foster real public places – environments where people enjoy the company of strangers.

John Morris Dixon, FAIA

Preface

Richard M. Rosan,
FAIA
President,
Urban Land Institute

Creating competitive, attractive cities that are cherished for generations is the goal of many cities, developers, planners, and urban designers today. Anyone who works in any aspect of the land use profession affects where and how people live, work, and play. And, while there is much left to be accomplished, people around the world are rediscovering the power of cities to connect people, to give them a sense of pride and belonging.

It is becoming increasingly obvious that as our world becomes more urbanized, cities are struggling with many of the same issues: managing growth, responding to demographic changes, building enough affordable housing, providing adequate transportation options and parking, and in general, finding the best way to rebuild, restore, and renew our urban areas.

The impact of urbanization is evident in countries around the world. People are moving to cities and urban regions as never before. By 2025, the United Nations projects that urban population growth will make up about 90 percent of the world's population growth. It estimates that in 20 years, 85 percent of the population in the United States will live in urban areas; in Europe, 83 percent; in Asia, 55 percent; and Africa, 54 percent.

Even at a 55 percent urbanization rate, Asia's population is so huge that an enormous number of people—well over 1 billion—will be living in urban areas. Currently, China, with 1.3 billion people; and India, with close to 1.1 billion, house more than one-third of the world's population. The U.S. ranks a distant third, with 295 million; followed by Indonesia with 241 million; and Brazil, with 186 million.

We can expect mega cities—a term coined by the UN to describe cities with at least 10 million inhabitants—to become increasingly commonplace. The United Nations projects that in just 10 years more than 20 cities will have more than 10 million people. Of these, only two—New York and Los Angeles—will be in North America. Six cities—four in Asia, two in Latin America--will have populations exceeding 20 million.

Clearly, it's not a matter of whether growth will occur. It's how and where growth will occur. To be sure, growth brings economic and social benefits, but if growth is mismanaged, it also can mean greater poverty, inadequate infrastructure, land scarcity, and a deteriorating environment.

Building competitive cities means building more than just places to live and work. It's about creating places that inspire, places with character, places that draw people through a powerful sense of identity. In cities around the world, changing demographics and changing household formations are having a profound effect on what is built and where it is built.

While every city has its own personality, there seem to be common characteristics behind every successful urban regeneration: strong political leadership; the creation of an environment for intellectual stimulation and creativity—to be a "brain-gain" city rather than a "brain-drain" city; tolerance of diversity; a commitment to provide housing to people with a variety of incomes; a solid track record in creating long-lasting public-private partnerships; a commitment to transit-oriented development and transportation infrastructure; and a dedication by the local officials to aggressively preserve land for parks and open space.

The cities that are able to offer a high quality of life—in the form of efficient transportation, recreational and cultural amenities, diverse neighborhoods, and a safe, clean, lively environment—will be the winners. This applies to both high-growth cities, which are scrambling to keep up with population increases, as well as low-growth cities, which are scrambling to retain and attract residents.

Creating places that give wonderful memories should be the ultimate purpose of urban regeneration. As community builders and place makers, all of us have a tremendous responsibility in shaping both the private and public space in which people carry out their lives.

180° Design Studio

1656 Washington
Suite 270
Kansas City, MO 64108
816.531.9695
816.531.9695 (Fax)
Kklinkenberg@180deg.com
www.180deg.com

180° Design Studio

180° Design Studio   New Town Theater District
                             St. Charles, Missouri

The distinctive crescent shape of the two-city-block site called for an appropriately formal design response for this mixed-use development. The symmetrical pair of curved buildings, each with about 28,400 square feet, will house residential units over street-level retail, civic, and office spaces. At the far ends will be matched loft buildings of about 10,800 square feet each. The central pair of pavilions will be linked by an over-the-street bridge to form one office building of about 27,400 square feet. Delicate canopies along the curved buildings are patterned after colonnades from the Old Market in Omaha. Smaller two-story "carriage house" residential structures behind the formally laid out buildings will make a scale transition to adjoining neighborhoods. The Classical design of the complex will create "an instant landmark," in the words of planner Andres Duany.

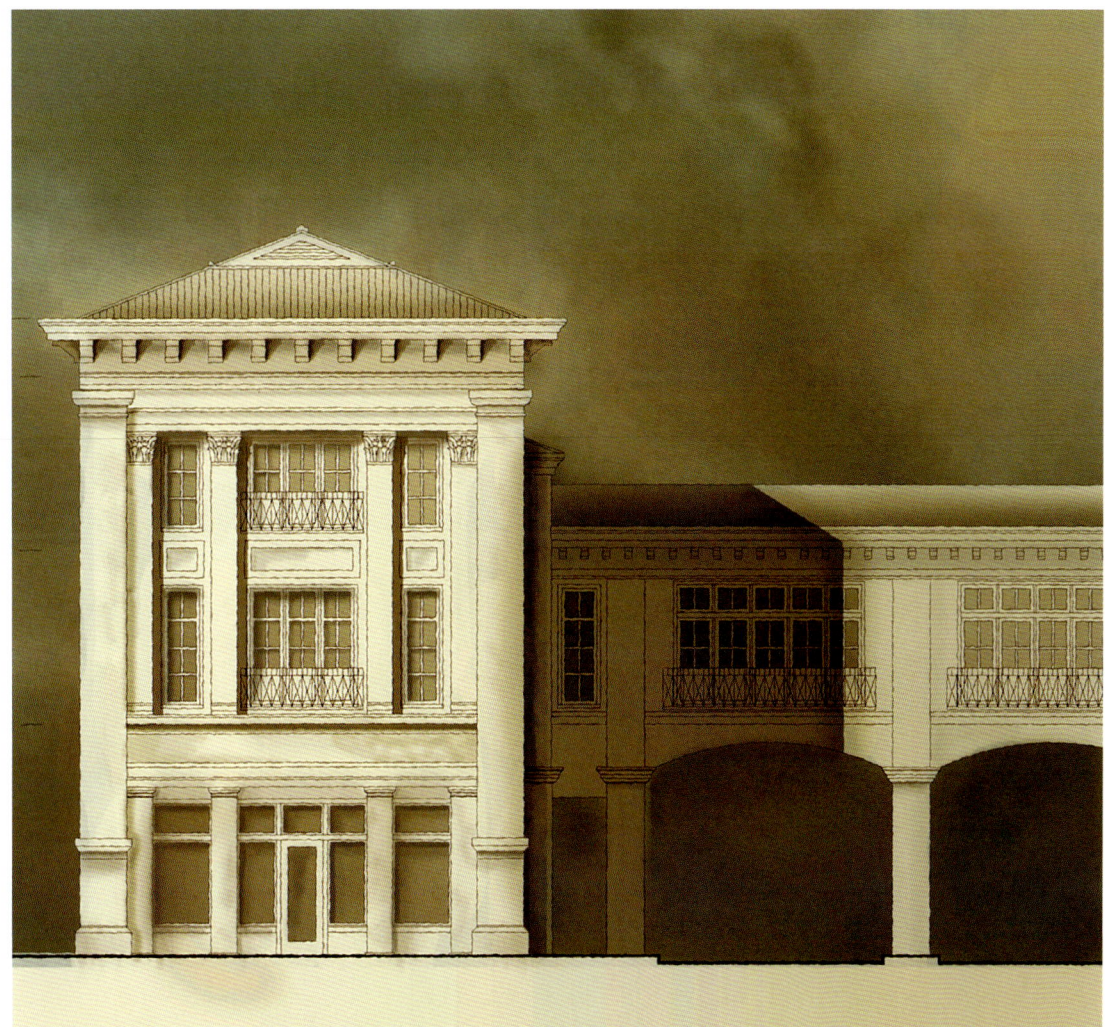

Opposite, top: Overall view of crescent around public plaza.

Opposite, bottom: Portion of crescent with steel canopies of historical character.

Above left: Central pavilions linked by bridge over street.

Left: Elevation of one central pavilion.

Rendering: Arnold Imaging.

180° Design Studio

New Longview
Lee's Summit, Missouri

The site of this 260-acre planned community is Longview Farm, the 90-year-old country estate of a lumber baron, which is listed on the Historic Register. The design preserves 14 of the 17 farm structures and integrates them into a walkable environment. Numerous previous proposals for the property had been turned down by the community. A multi-day public design charrette conducted with the community achieved almost unanimous support for this master plan. Months of work were spent with city staff on the details of the rezoning. The mixed-use development will include 1,100 residential units, 250,000 square feet of retail, and 250,000 square feet of offices, plus space for civic structures and public open spaces totaling 70 acres. A regional arterial road that was planned to bifurcate the site was redesigned as a multilane boulevard that accommodates the traffic while enhancing the community.

Above left: Master plan, with existing mansion at lower right.

Left: Park land around mansion.

Clockwise from photo above: completed houses; residential neighborhood; civic center reusing farm buildings; multifamily crescent; variety of building types along boulevard; theater; houses and neighborhood retail.

180° Design Studio   Crescent Creek
                               Raytown, Missouri

For a 22-acre site adjoining an existing Post-World-War-II neighborhood, planning and architectural design were carried out for 130 residential units. It was essential to make the development attractive yet very affordable. Rezoning for a planned residential district required establishment of strict design regulations. The six types of dwellings include townhouses and single-family houses, ranging in size from 1,100 to 2,600 square feet and in price from $150,000 to $275,000. A pool and clubhouse are included in the project. Existing streets were extended through the new development but "tamed" to reduce speed and discourage through traffic.

Above: Typical street and houses.
Left: Master plan.
Below left: Central green.
Bottom left and below: Typical houses nearing completion.

180° Design Studio

Ottawa University Master Plan
Ottawa, Kansas

This campus master plan features a new student learning center and upgraded residential and athletic facilities. The objective was to create a stronger campus feeling, with a true center and a series of "outdoor rooms." Parking is dispersed around the perimeter of the site. An axial green at the main entrance offers numerous parking spaces integrated into a formally landscaped setting. Residential buildings line another axial green. A tower marking the center of the campus rises above a circular plaza at its front and aligns on the far side with the 50-yard-line of the athletic field. New buildings are to vary in character from relatively large-scaled and symmetrical for focal buildings to more intimate and irregular for the residential structures, all of them clad in traditional materials.

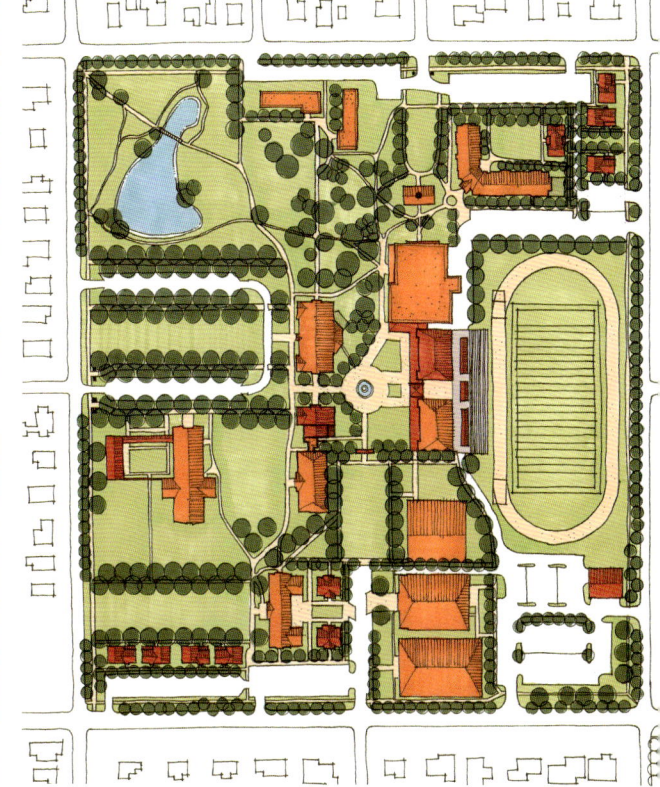

Top: Circular plaza and signature tower at campus center.

Above: Residential green, with small-scaled student housing at right.

Above right: Master plan.

Right: Axial entrance green.

180° Design Studio — Residential Projects

Top and above left: Elevation and digital model of 71st and Metcalf project.
Above: Aerial rendering of Union Hill development.
Left: Beachtown Galveston Village model home.
Below: Longfellow Court residential buildings.

Several residential projects show the firm's adaptation of traditional housing forms to specific circumstances. The 71st and Metcalf development in Overland Park, Kansas, proposes 24 townhouse and flat units, ranging from 650 to 2,000 square feet to serve different markets, on a 2.1-acre suburban infill site. Union Hill, Phase 3, in Kansas City includes apartments, townhouses, and retail on a two-city-block site complicated by steep topography. A single-family model home for Beachtown Galveston Village is meant to establish a design image for the community, dealing effectively with restrictions such as the raised first floor required at this beachfront location. Longfellow Court in Kansas City, developed by the firm itself, proposes 18 units, including detached single-family, townhouses, duplexes, and flats, in a scheme that achieved unanimous approval for 35 zoning variances.

Altoon + Porter Architects LLP

444 South Flower Street
48th Floor
Los Angeles, CA 90071-2901
213.225.1900
213.225.1901 (Fax)
apa@altoonporter.com
www.altoonporter.com

Altoon + Porter Architects LLP

Knox Shopping Centre
Melbourne, Australia

As the first phase of a long-range plan for the 500-acre Knox City development, this 200-acre Shopping Centre contributes to a larger urban framework, incorporating a mix of retail, civic, and residential buildings, including existing mixed use buildings. Establishing linkages to wetlands and transportation, the overall plan also provides for sports, educational, and light industrial facilities. The Shopping Centre turns a familiar suburban retail form into an urban environment of streets and squares, with restaurants and cinema. Six separate precincts, each with a memorable name and visual personality, ease wayfinding within the centre and offer distinct merchandising opportunities. Calling upon the rich pictorial and metaphorical vocabularies provided by the site between Melbourne and the Dandenong Ranges, the design evokes the Australian life style in the expression of native timbers and lodge forms, in the palette and patterns of the vineyards, in the broad verandahs and lawns of the town square, and in the bright lights of the urban scene. As a result, visitors enjoy a marked sense of place that is as diverse as the region itself.

Above: Two details evoking regional traditions of broad overhangs, industrial materials, and angular timbers.

Right: Traditional cityscape elements of urban square and pedestrian paseo.

Above: Intimate pedestrian passage opening into broader square.

Right: Activity along passage sheltered by tiers of canopies.

Far right: Retail interior.

Photography: Stuart Curnow.

Altoon + Porter Architects LLP

Victoria Gardens
Rancho Cucamonga, California

In a dramatic departure for a retail development, Victoria Gardens is conceived not as a project but as a town center reflecting the values of a multigenerational community. The design of the 1.2-million-square-foot development is rooted in the elements common to all towns and in the special qualities of this singular place. The plan is organized around a grid of streets, a town square, courtyards, paseos, pocket parks, and plazas. The individuality of shops and buildings thrives on the underlying sense of urban order. Two-story shops, second- and third-floor offices, and residential lofts above retail all contribute to the character and scale of a cityscape. Civic uses such as a children's performing arts theater, a conference center, and a central library around a public square all add a civic quality. The development occupies 12 city blocks at the heart of a 160-acre planned community and includes three major department stores, a cinema, and 150 specialty shops. The presence of residential development contiguous to Victoria Gardens reinforces its town-center role.

Top: Hi-tech building detail.

Above: Sleek Modernist loggia with characteristic street furniture.

Right: Cityscape recalling traditional towns.

Above right: Food Hall and other structures evoking old industrial districts.

Opposite, bottom left: Structure in Spanish Colonial spirit.

Opposite, bottom right: Interior of Food Hall.

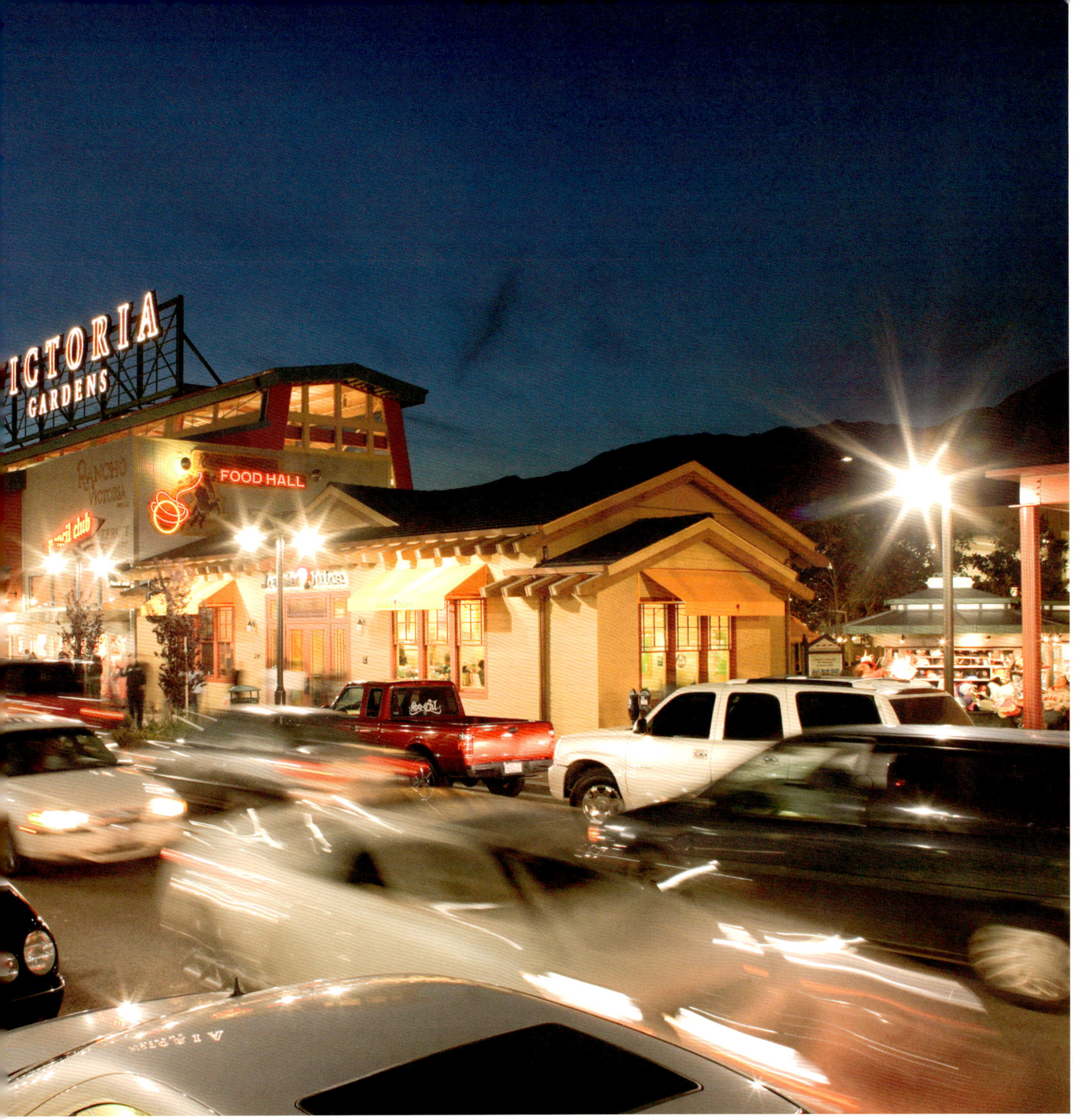
ICTORIA
GARDENS
FOOD HALL

california pizza kitchen

RANCHO
VICTORIA
CAFES

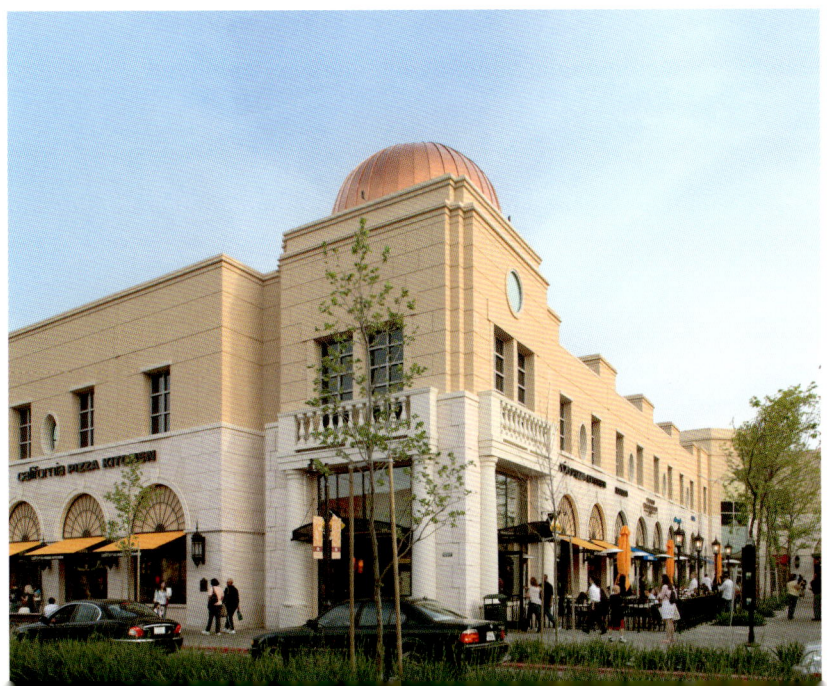

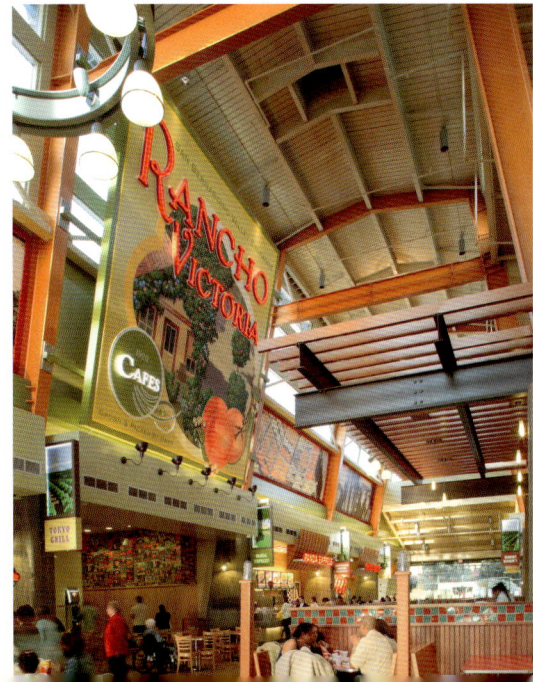

Altoon + Porter Architects LLP

Fashion Show
Las Vegas, Nevada

A 20-year-old, well-performing but barely noticed retail center has been transformed into one of the most memorable landmarks on the Strip, a shopping environment that could hardly be replicated anywhere in the world. The complex has been doubled in size – without interrupting its retail operations – to 1,785,000 square feet, with eight department stores vs. the earlier five, and totally reconfigured to capitalize on the evolving pedestrian nature of the Strip. The enlarged center establishes its unique presence with a 600-foot-long high-tech canopy known as The Cloud. Suspended 180 feet above a 72,000-square-foot plaza, The Cloud provides visual entertainment as well as welcome shade. A sophisticated audiovisual system projects images to and from the plaza and The Cloud with a series of super-sized LED screens. Fashion shows from around the world and within the center are projected as they take place in real time, thus delivering a vivid embodiment of the project's name. Sophisticated controls allow projections of other world events, as well, in real time. Inside, a great hall,

Top: Whirling geometries of canopy and building.

Above: Projections on round screens in great hall.

Photography: Erhard Pfeiffer.

Above: Canopy known as The Cloud above plaza at center's entrance.

Right: Actual fashion show in great hall.

Far Right: Great hall's elevated, theatrically lighted runway/stage.

850 feet long and 150 feet wide, gives visitors a chance to watch real fashion shows on an elevated runway/stage. The complex now includes 4,800 parking spaces. It is bound to be on the "must-see" list of every visitor to Las Vegas.

Above: Canopy as unique evening landmark.

Left: Ample interior circulation areas.

Below: Dynamic building corner.

Photography: Erhard Pfeiffer.

annex|5
An Epstein Design Group

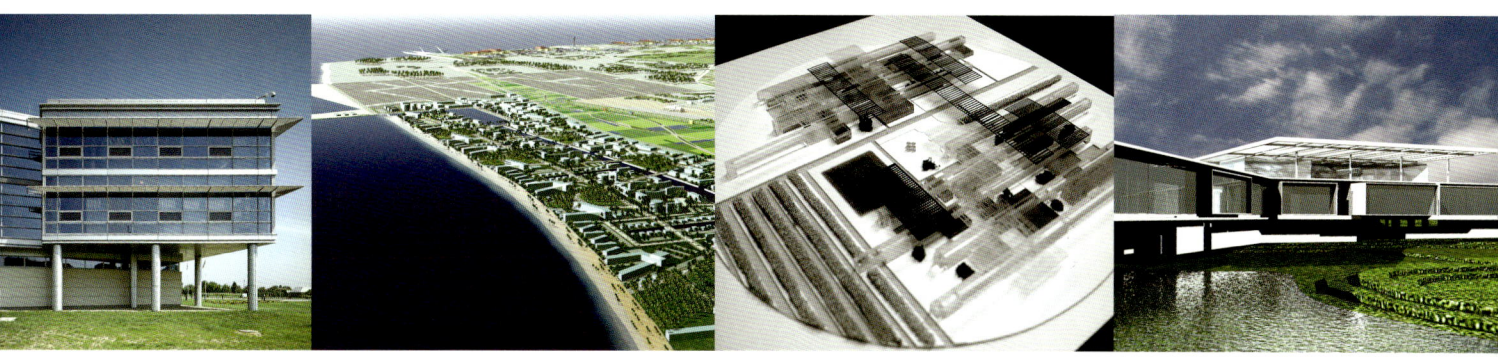

600 West Fulton
Chicago, IL 60661
312.454.9100
312.429.8175
312.559.1217 (Fax)
www.annex5.net
ametter@annex5.net

New York
Tel Aviv
Warsaw
Beijing
Shenzhen
Los Angeles
San Antonio

annex|5

annex|5
An Epstein Design Group

Motorola Global Software Group
Krakow, Poland

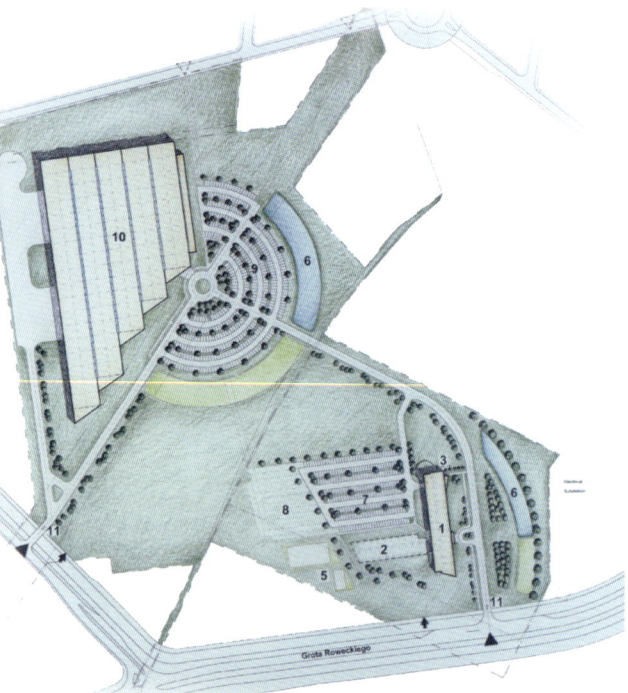

Left: Site plan with completed building and Phase II wing at lower right, proposed manufacturing facility at upper left.

Middle left: Entry detail.

Bottom: Long façade.

Facing page top: Building at night.

Facing page middle: Façade detail.

Facing page bottom: Narrow front facing street.

Photography: Christian Richters.

Intended for the development of proprietary software, this facility is located on the outskirts of Krakow as part of a master planned technology campus associated with the 800-year-old Jagiellonian University – Poland's equivalent to Oxford. Situated among high-profile institutions such as the nearby Papal Academy of Theology, this first private-enterprise building on the campus houses research and development offices, computer labs, and a cafeteria. The design addressed its relationship to these other complexes through its siting strategies and its external expression as an abstract, industrialized volume – in effect a metaphor of software development. The three-story, 5,500-square-meter (59,400-square-foot) structure takes the form of a narrow linear volume oriented perpendicular to the road, yet it provides a frontal reading as approached from the city. The narrow floor plate affords all employees access to daylight, views, and natural ventilation. Workstations are held back from the floor-to-ceiling, double-walled glass envelope to improve circulation and establish a non-hierarchical interior. Amenity spaces and interactive zones are highlighted through shape and location. Phase II will add a 5,000-square-meter volume at right angles to the initial one and parallel to the street, creating a screened car park.

annex|5
An Epstein Design Group

Planning Design of Haikou West Coast
Haikou City, China

Above: Composite land-use plan.

Left: Detailed master plan of area at left in land-use plan.

Below: Aerial image of area in this master plan.

Awarded First Prize in an international design competition, this plan envisions environmentally responsible development along 26 kilometers of shoreline in Hainan province, an island just off the southeast China coast. Haikou is the political, economic, and cultural center of this tropical island. Considering the expressed goal of developing Hainan into a national four-season garden and holiday resort, Haikou has a unique opportunity to become an ecological showcase. While Haikou has seen tremendous growth, much of it has been unplanned, uncompleted, or unsuccessful. This master plan outlines a framework within which Haikou could implement a successful smart-growth strategy. It defines areas of medium- and high-density residential development, public open spaces, and resort accommodations, with creative architectural design guidelines. By properly harnessing the area's resources, the plan could make Haikou world-famous as a green international resort and metropolis.

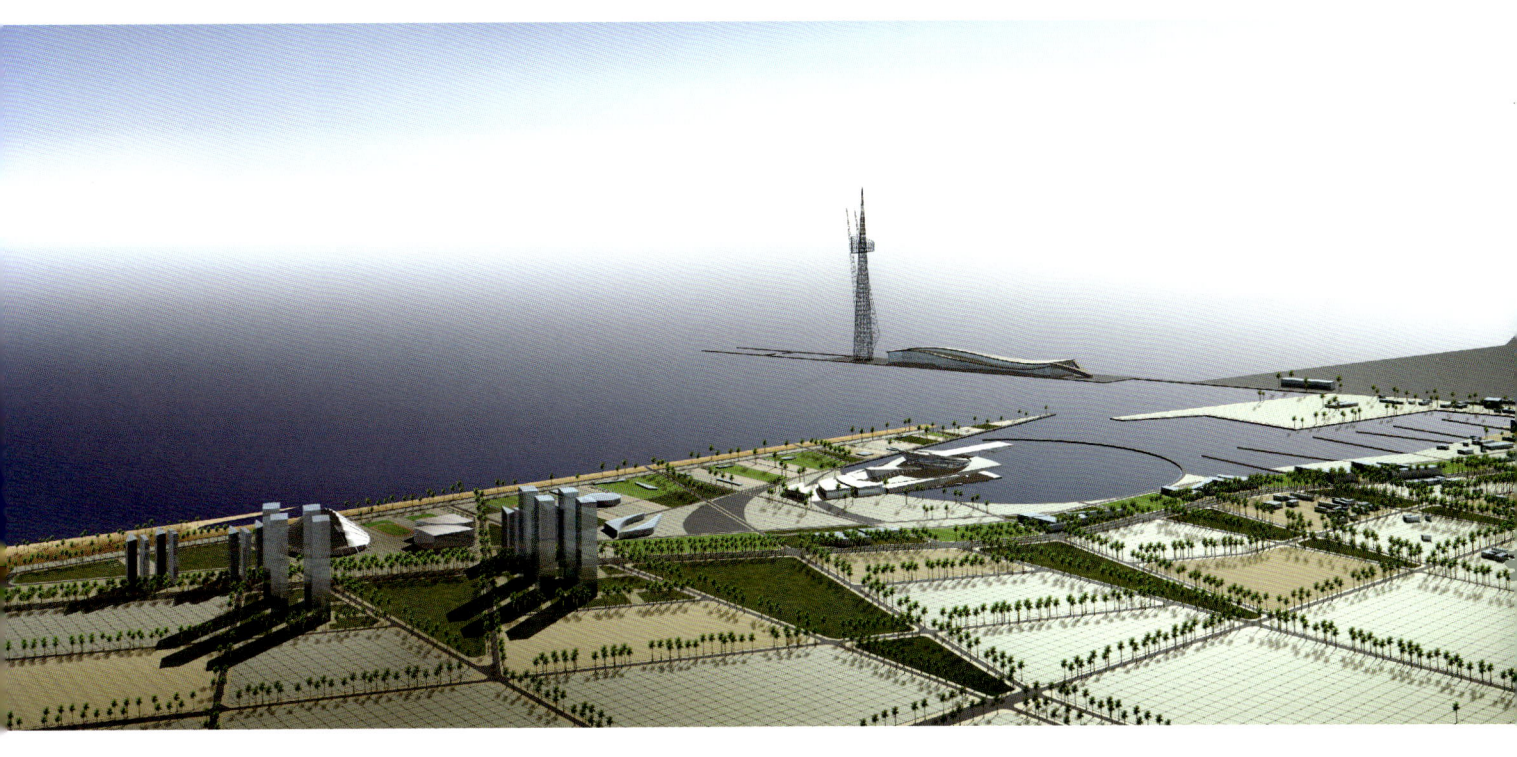

Above: Bird's-eye view of area at right in land-use plan.

Below left: Concept image of "garden wall" residential units.

Below right: Concept image of beach villas.

annex|5
An Epstein Design Group

Serta International Center
Hoffman Estates, Illinois

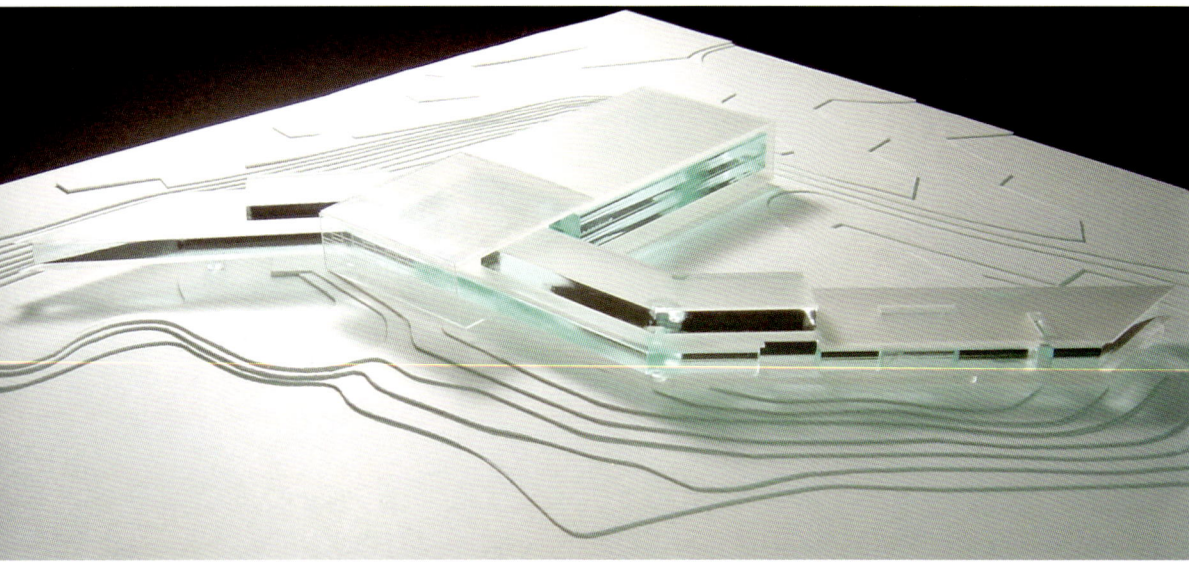

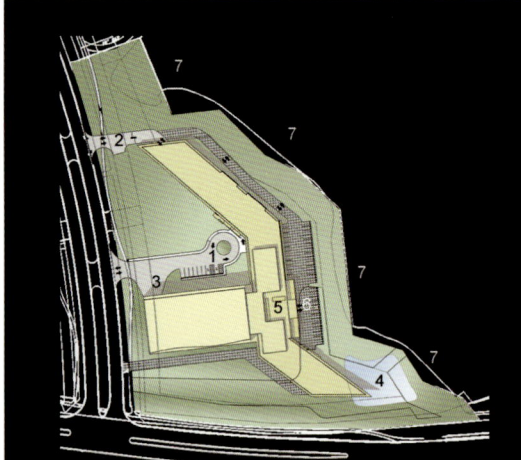

Top and bottom: Model of building in landscape.

Middle left Site plan: 1 main entry drop-off; **2** employee parking entry; **3** service dock; **4** retention pond; **5** outdoor deck; **6** grass paver/fire lane; **7** protected wetlands.

Facing page top: Rendering as seen from grasslands.

Facing page middle: Projection over retention pond.

Facing page bottom: Shading studies.

Model photo: Andrew Metter.

The 80,000-square-foot headquarters of the Serta International Center will be located on a seven-acre parcel in an environmentally sensitive business park. The site immediately adjoins a protected wetland that is essential for storm water management. All parking is located under the building, reducing impervious surfaces on the site and generating the 65-foot width of the office structure. The building appears to hover above a sea of grass, its glass walls shaded by horizontal louvers echoing the prairie landscape. Its massing reflects its two main functions: office space in the "bar" portion along the edge of the wetlands and research and development in a "box" volume. These two join at the center, where public spaces such as showrooms, lunch room, and training auditorium are located. The second level opens to a deck area covered by a canopy with a woven pattern that reflects Serta's interest in textures and fabrics.

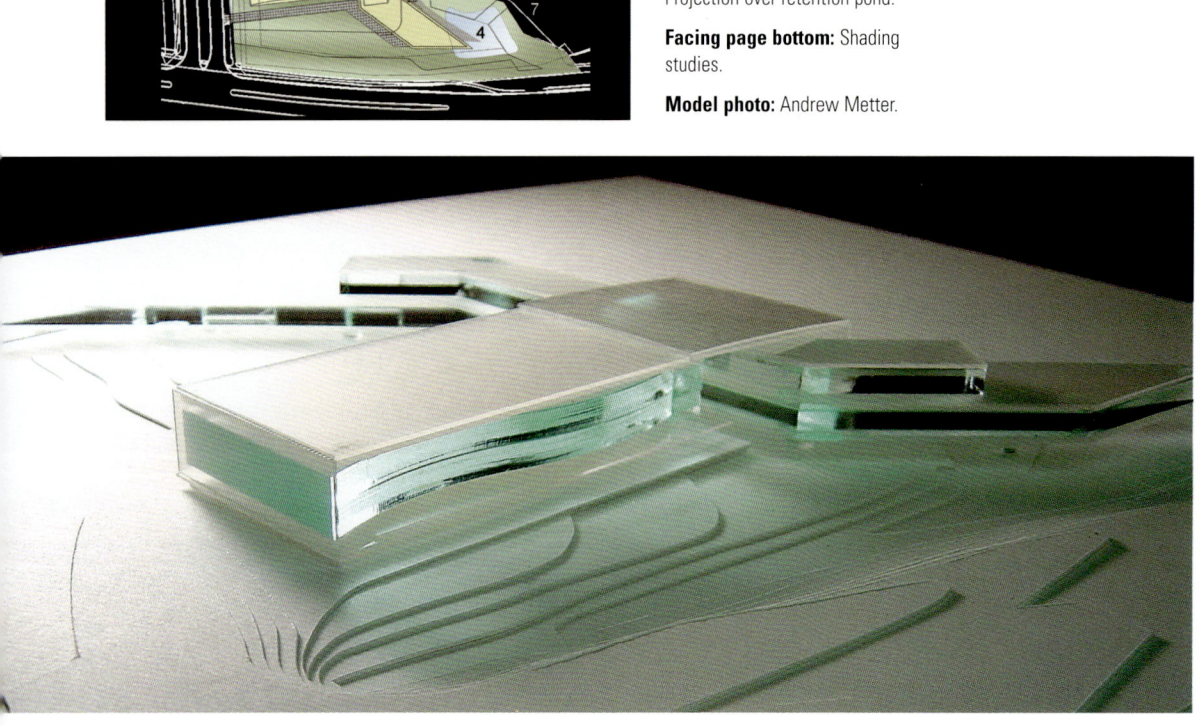

annex|5
An Epstein Design Group

Brighton Village
Coolidge, Arizona

Located 53 miles from Phoenix, Brighton Village is a new planned community of 1,500 acres, projected to house a population of 25,000. Annex/5 was commissioned to design the 57-acre Town Center, which will include civic, small-scaled retail, and higher-density residential functions, along with support functions such as day care, college outreach, and YMCA facilities. A proposed 350,000 square feet of buildings will be organized for walkability, climatic response, and water management, using graywater reservoirs, water courts, and cisterns. One of the major elements affecting the plan is a large "arroyo" channel, which will double as a greenbelt and open up views of distant mountains. The plan proposes community gardens and farmer's market stalls. The plan is based on solar orientation, composed of a grid of shaded pedestrian arcades oriented on a north-south axis. An additional shading layer of horizontal trellises allows for the generous use of glass, resulting in an intimate connection to the landscape.

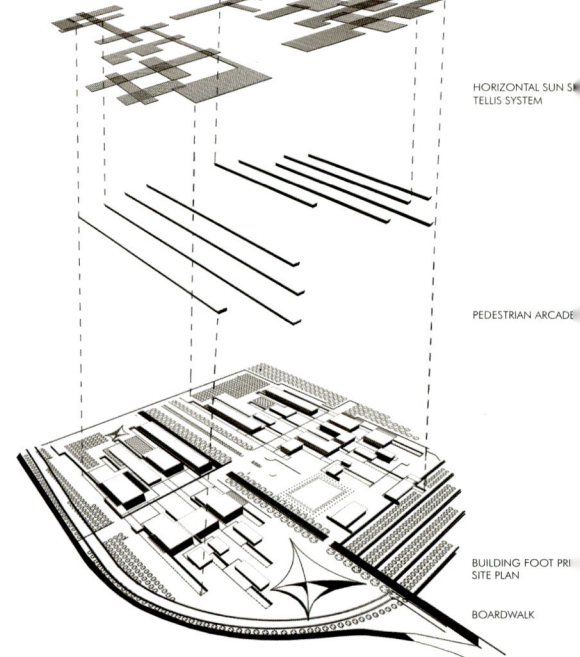

TOWN CENTER COMPONENTS

Top: Kit-of-parts concept image.
Middle right: Assembly of kit of parts into town center.
Bottom right: Model of town center.
Left: Plan of town center, including school/athletics campus.

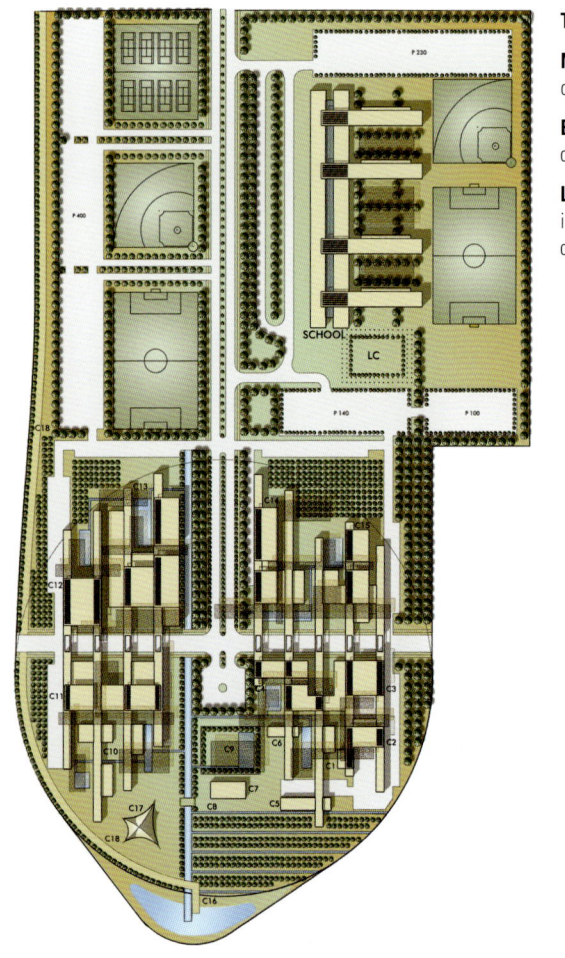

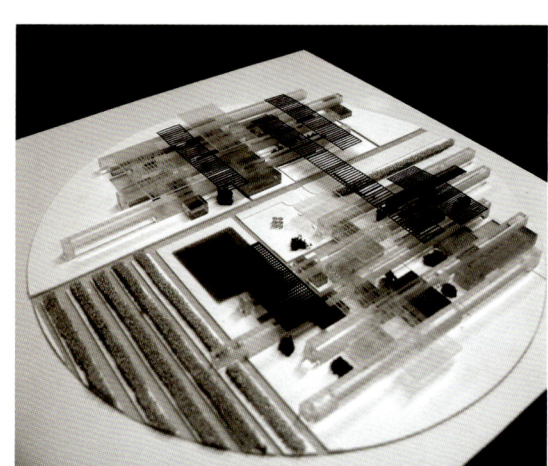

32

Austin Veum Robbins Partners

One America Plaza
600 West Broadway
Suite 200
San Diego, CA 92101
619.231.1960
619.231.1920 (Fax)

info@avrp.com
www.AVRP.com

Los Angeles
213.627.7170

Ventura
805.652.1129

Tijuana, Mexico
011.52.664.686.3986

**Austin
Veum
Robbins
Partners**

Allegro Tower Apartments
San Diego, California

Above and right: Tower seen from harbor.

Top right: Details of exterior and balconies.

Above right: Eighth-floor pool deck.

Opposite: Tower seen from transit line.

Photography: David Hewitt and Anne Garrison Architectural Photography.

This 204,000-square-foot project was shaped in part by its impact on neighboring buildings. Much of the site is in parts of the Little Italy District where high-rise buildings are not allowed. In response to zoning restrictions and considerations of neighborhood context, the project's 201 apartments are divided into a slim tower which is flanked by an 8-story block on one side and a 5-story block on the other. By making the tower very narrow in its east-west direction, AVRP minimized its shadow on surrounding structures and streets. The tower floor plan is single-loaded, so that every unit faces the bay. Retail spaces at the street level help to activate the street, and there are four levels of parking below ground level. An eighth-floor pool deck offers ample natural light and views of the neighborhood city skyline bay. The cast-in-place concrete structure is clad partly in cast-in-place concrete and partly in GFRC panels.

**Austin
Veum
Robbins
Partners**

Egyptian Lofts
San Diego, California

Bush's historic Egyptian Theater is embedded in the heart of this mixed-use complex, scheduled for late 2005 completion. The aim of the design is to acknowledge the Egyptian Revival style of the theater's colonnaded entrance front and link it to the modern housing constructed around it. Dating from 1926 and originally serving as the foyer for the theater, the monumental entry is being fully renovated, with repairs to deteriorated features and reconstruction of missing but documented elements. The foyer will have the same appearance but play a different role as outdoor seating for a restaurant or café. The new surrounding structure includes retail at the street level, with 80 one- and two-bedroom condominium units on the upper floors. Defining architectural features of the residential construction allude to Egyptian style in the use of Art Deco elements, such as occasional round windows and planar projections that cast deep shadow lines. An angular fin reaching above the flat roof, reminiscent of the "blade" signs of historic movie houses, identifies the complex.

Above: Long facade, with niche at historic theater.

Right: Building corner, showing relationship to existing neighbor.

Photography: David Hewitt and Anne Garrison Architectural Photography.

Rendering: Jason Brown.

Left: Partial elevation view, with vertical fin sign.

Right: Restored Egyptian Theater façade between mixed-use wings.

**Austin
Veum
Robbins
Partners**

Park Laurel on the Prado
San Diego, California

Prominently sited along a broad approach to Balboa Park, this luxury complex includes two similar condominium towers turned at right angles to each other maximizing views. Each building is designed with four condominiums per floor, and condo elevators opening directly into units, creating a 90-percent efficiency above the second floor. Two-story penthouses crown the buildings. The lower two floors contain retail and office space, including an existing bank that has been moved to a new location within the complex. A canopied and vaulted entrance pavilion differentiates the residential entrance from the lower-floor commercial uses and recalls the landmark lattice house in Balboa Park.

Above: Rendering of two towers seen from boulevard.

Left: The completed tower, with entrance in foreground.

Below left: Tower two rising from the two-story commercial base.

Photography: AVRP.

Rendering: Laurel Watts.

**Austin
Veum
Robbins
Partners**

Smart Corner
San Diego, California

Located in San Diego's East Village, Smart Corner is intended as an ideal "smart growth" project. It includes a 19-story condominium tower, a 5-story office building, retail at the street level, and a trolley stop located between its two buildings. Altogether, the project includes 298,000 square feet of residential and 111,500 square feet of office space, plus four levels of underground parking. The project's public open space, including the trolley stop, amounts to 31,000 square feet. The basic 450-square-foot studio unit, starting at $185,000, is designed as flexible space or "morphable rooms," which can be divided by furniture, counters, or partitions. They are adaptable to Murphy beds or partial lofts over kitchens. The developers have set aside 25 units for buyers whose incomes qualify them for special mortgage terms under a first-time-owner program.

Top left: Complex's two buildings, with trolley stop in public space between.

Left: Residential building, showing two distinct architectural volumes.

Above right: Angular walls along trolley line.

Austin Veum Robbins Partners

The Pinnacle Museum Tower
San Diego, California

This upmarket residential condominium is intended to share its Marina District site with the San Diego Children's Museum. A low-rise podium with retail spaces fronts onto three streets with office space above. A single tall tower was chosen in lieu of a lower and bulkier building to minimize obstruction of views from neighboring buildings. The 182 residences are accommodated in a slender 434-foot tower with open space to the south and southeast. The tower and its podium total 350,000 square feet of mostly residential space and 10,000 square feet of pool deck over retail spaces. There are three levels of parking below ground for residents and for the Children's Museum staff and visitors. The Children's Museum will total 32,000 square feet above ground plus basement level. The tower is located along the Harbor Boulevard trolley line and the Martin Luther King Promenade. It is also a short walk from historic and rejuvenated downtown destinations. The project is scheduled for completion in the fall of 2005. The incorporation of the Children's Museum and substantial open space adjacent to the development promises to promote a sense of community and enhance the environment for children and adults alike.

Above: Pinnacle tower rising over public open space.

Left: Broad view of tower in urban setting.

Far left: Tower with Children's Museum at its base.

Photography: AVRP.

Beeler Guest Owens Architects, L.P.

4245 North Central Expressway
Suite 300
Dallas, TX 75205
214.520.8878
214.520.8879 (Fax)
bgoemail@bgoarchitects.com
www.bgoarchitects.com

Beeler Guest Owens Architects, L.P.

The Highlands of Lombard
Lombard, Illinois

Located in a southwest suburb of Chicago, the Highlands is a high-density infill project with a total of 403 units. A change of elevation of 35 feet across the five-acre site posed design challenges. A five-story, seven-level scheme, with a unique combination of entries, stairs, and elevators, achieved the goal of 81 units per acre. At the center of the development, a six-and-a-half story parking structure is concealed by the surrounding residential buildings. Residents have direct-access parking on the levels where they live. Variations in the design of facades and the elevations of entries reduce the visual impact of this large-scale project on its neighborhood. Construction with light-gauge noncombustible steel turned out to be faster, more accurate, and safer than conventional wood framing. The precast parking garage was significantly faster to erect than cast-in-place construction. The Highlands is a success for the owners, is appreciated by residents, and is well received by the surrounding community.

Above: Typical street elevation.

Top right: Pool courtyard.

Above right: Interior public space.

Facing page: Exterior, showing variety of architectural features.

Photographs: Scott Brennan/Hedrich Blessing.

Beeler Guest Owens Architects, L.P.

5225 Maple Avenue
Dallas, Texas

Left: Site plan.

Below left: Entry tower on main façade.

Facing page top: Lobby space and pool courtyard.

Below right: Details of pool deck and building exteriors.

Photography: Mark Guest & Jeffrey Massey.

Just a stone's throw from the center of Dallas, the Maple Avenue infill project provides luxury housing for young professionals who work downtown and in nearby hospitals. The complex fits 224 units and two parking structures on its 4.9-acre site, with a design aesthetic that borrows from the area's light industrial uses of wood, concrete, and steel. A monumental four-story steel-and-glass tower, along with two large arched automobile gates, creates a dramatic entry statement on the Maple Avenue façade. An interior courtyard with a pool and pavilion makes reference to the same rugged aesthetic. The interiors also play variations on the industrial theme, with dark-stained concrete floors, brick walls, and exposed steel structural elements. The project sets an instructive example for a neighborhood undergoing growth and change

Beeler Guest Owens Architects, L.P.

Easton
Dallas, Texas

Situated in a neighborhood first developed in the 1920s and 1930s, primarily with Craftsman-style houses, this three-story development is designed to focus attention on its compatible details and thus minimize its apparent bulk. Seen from the street, the highly articulated elevations divide the building into portions of single-family scale. The three-acre, 150-unit project is the latest infill effort in a 10-year-old process that is revitalizing older residential areas north of downtown along the North Central Expressway. The dwelling units surround a central precast parking garage, effectively blocking its view for neighbors. Also hidden from the community are the project's active pool courtyard and a passive courtyard with an outdoor fireplace. Inside the complex are luxurious interiors with many amenities. The development targets the growing population of young, single professionals who wish to be connected to the single-family lifestyle of the northern suburbs as well as the high-energy activities of downtown Dallas.

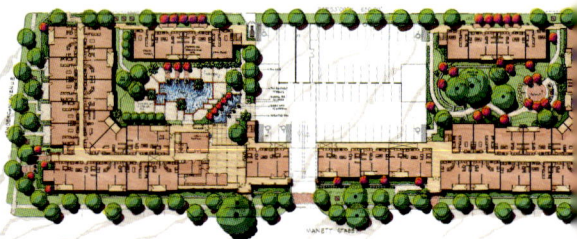

Above left: Craftsman-related details along street.

Left: Pool courtyard.

Above: Plan of Easton Development Site.

Photography: Mark Guest & Darren Dobbins.

Beeler Guest Owens Architects, L.P. The Davis Building
Dallas, Texas

Once a proud symbol of the Dallas financial district, this building has been adapted to house 183 distinctive residential lofts. Originally the headquarters of Republic Bank, the structure played a prominent role in the emergence of Dallas as a major city. It had been unoccupied for over a decade, vandalized and deteriorating. In order to qualify for historic tax credits the renovation had to meet the strict standards of the National Park Service. The individual units are designed as "hard lofts," revealing original building elements such as concrete framing and exposed ducts. Juxtaposed to them are new maple doors and cabinets, granite counters, and oversized oval bathtubs. The penthouse that includes the cupola has five interior levels, two private terraces, and a glass-bottomed hot tub.

Above left: Loft unit.
Above: Internal air and light shaft.
Below left: Building in cityscape.
Below: Cupola and penthouse terrace.
Photography: Mark Guest.

Beeler Guest Owens Architects, L.P.

Union Station
Union, New Jersey

The first transportation-oriented development (TOD) in the state of New Jersey, Union Station will have a 20-minute link to New York City via a metro rail station bordering the property. Its 227 corporate suites will provide limited views of the city skyline. Although surrounded by a residential community, the project draws architectural inspiration from the industrial and shipping facilities that have traditionally occupied the New Jersey shore of the Hudson River. The complex has one level of underground parking, with four levels of wood-framed construction rising from this concrete podium. On top of the garage is an open courtyard with a pool, water features, a fire pit, and private patios adjoining first-floor units. Amenities characteristic of a luxury hotel include a business center, a 50-seat theater, a cybercafé, a fully equipped and staffed fitness center, and a Great Hall for entertaining guests. To tenants, the complex offers a variety of furnished or unfurnished units, available by the week, by the month, or on one-year leases.

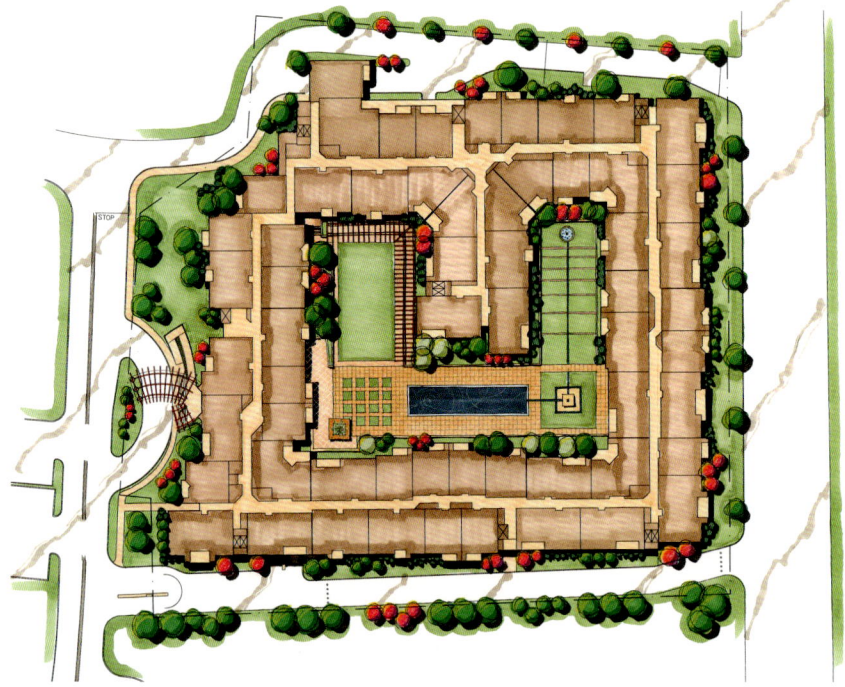

Top: Site plan.
Above: Closer view, showing architectural details.
Left: Overall view.

Callison Architecture, Inc.

1420 Fifth Avenue
Suite 2400
Seattle, WA 98101
206.623.4646
206.623.4625
info@callison.com
www.callison.com

Callison Architecture, Inc.

Callison Architecture, Inc.

Ayala Center Greenbelt
Makati City, Philippines

Above and left: Retail pavilions around water garden.

Below: Entertainment and nightlife portion of project.

Facing page: Interpenetration of building and landscape.

Photography: Chris Eden.

Greenbelt is a four-story open-air pavilion located within the Ayala Center mixed-use complex in Manila's central business district. While enclosed shopping centers are the rule in this tropical climate, this complex of buildings is interwoven with nature, providing a welcome oasis in a dense city. Louvers, canopies, overhangs, and ample ventilation protect patrons from the weather. The project fits around existing features – a park, a church, and a museum – and was laid out to save 440 mature trees. A "garden wall" concept allows the complex to meet a variety of needs. Its city side presents the unified street front that more fashion oriented retailers desire, and on the other side the park virtually grows into the terraced building, with opportunities for outdoor dining. The project's series of four-story pavilions offers a spectrum of environments, starting next to the museum with quiet bookstores and galleries, proceeding to home and fashion oriented tenants and terrace restaurants near an active intersection, then to a lively zone of music and video stores with dining, cinema, and nightlife.

Callison Architecture, Inc.

Grand Gateway
Shanghai, China

Adjacent to the Xuijahui subway station, one of Shanghai's major transportation nodes, Grand Gateway comprises a 1.1-million-square-foot, seven-level retail podium, two 34-story residential towers, two 52-story office towers and a nine-story service apartment. One of the first comprehensive collections of Western-style retailing in Shanghai, the shopping center introduces a variety of themed environments: an outdoor street of restaurants and sidewalk cafes modeled after those of Paris, a lively entertainment zone with Hollywood imagery, and shopping districts ranging from Chinese zones to fashion-oriented international areas. The design creates a synergy among uses for greater identity and impact. Connections to the subway pass through the shopping area to encourage retail volume; apartments are located near the outdoor dining zone as a convenience for residents and a benefit to restaurant operators; the entertainment component reinforces the shopping center and is an amenity for the service apartments and residential towers. The project was designed for flexibility to add or expand components as demand for such functions as offices and service apartments increases.

Above: Cylindrical retail atrium as city landmark and prominent entrance to project.
Top right: Interior of atrium.
Right: Towers exterior.
Below right: Multi-level shopping concourse.
Facing page: Pedestrian retail street within complex.
Photography: Chris Eden.

Callison Architecture, Inc.

Suwon Gateway Plaza
Suwon, Korea

As co-developer and anchor department store, Aekyung has become an industry leader pioneering a new type of retail destination for Korea. Suwon Gateway Plaza incorporates Aekyung at one end, a superstore/entertainment anchor at the other and specialty retailers that link them together, all built above one of the busiest transit centers in the country. Designed to enliven the traditional Korean shopping experience while capitalizing on the activity generated by the 30 million travelers who pass though the commuter station annually, Gateway Plaza represents the future of retail for the country. The 1,365,000-square-foot mixed-use complex also includes cultural facilities, restaurants, and office space. Responding to an increasingly competitive retail environment in a previously under-stored nation, Aekyung was the first in Korea to modify the traditional department store format by taking specialty retailers out of the store and into the mall. And while Korean department stores have traditionally been self-contained and indifferent to their settings, Gateway Plaza opens outward to the city and has become a signature feature of the city of Suwon. Integrating a unique merchandising mix and a strong architectural identity, the center is now being emulated across the country in numerous other developments and has set the standard for retail destinations.

Facing page: Main entrance to complex.

Above: Mixed-use development rising over existing transportation hub.

Far left and left: Interiors of retail mall.

Photography: Chris Eden.

Callison Architecture, Inc.

Metropolitan Tower
Seattle, Washington

Located at the edge of Seattle's retail core rather than in one of its established downtown neighborhoods, the Metropolitan uses hospitality-inspired design and service to attract residents to its uptown address. Flanked by office towers, upscale retail, restaurants, hotels and a new Federal courthouse, the location offers a mix of activities and uses. The 24-story, 386-unit luxury apartment tower rests on a seven-story podium that includes parking and streetfront retail. Its dramatic architectural profile encases 47 different unit floor plans and the grand hotel style lobby sets the tone for amenities including concierge and valet service, a full-service health club, business and conference center, guest suites and a spacious, eighth-floor garden terrace. The tower's plan and distinctive form are derived from the unique angular geometry of the site. The design objective was a sleek, sophisticated hotel ambiance that would fit the neighborhood image and the likely preferences of those who would live here. The exterior limestone base, precast panels, metal and glass echo the materials of neighboring buildings. The units stress elegant living accommodations and are laid out to take maximum advantage of spectacular views from cantilevered bays and recessed balconies. Residents ranging from their early 20s to mid-70s cite the array of services as well as proximity to social and work opportunities as reasons they chose to live at the Metropolitan. Despite its completion amid a sharp economic downturn, the building was nearly 90 percent leased when opened.

Above: Tower in its edge-of-downtown context.

Right: Glazed canopy providing pedestrian shelter.

Far right: Metal and glass details relate to surrounding non-residential buildings.

Photography: Chris Eden.

Canin Associates

500 Delaney Avenue
Orlando, FL 32801
407.422.4040
407.425.7427 (Fax)
design@canin.com
www.canin.com

Canin Associates

Canin Associates Grande Lakes Resorts
Orlando, Florida

A resort destination has been created here, knitting two hotels and the health club/spa structure between them into a complex with a distinctive ambience. But the elite Ritz Carlton and the less formal J.W. Marriott maintain their separate brand identities. The identity of the resort is established for arriving visitors by a mile-long road of picturesque landscaping designed to immerse the guest in the relaxed experience of a tropical resort. The amenities of the complex, in addition to the two hotels with their total of 1,600 rooms, include the health club and spa, two pools, and a golf course. One of the most memorable features of the 500-acre

grounds is the lazy river, a convoluted and meandering watercourse running between rustic stone walls and under lush plantings, along which guests can be propelled, playfully splashing or passively observing the setting.

Facing page: Pool and spa area.

Above: Lazy river passing under footbridge and fountain.

Above right: J.W. Marriott seen from lazy river.

Right: Master plan of J.W. Marriott (at left on plan) and Ritz Carlton properties.

Canin Associates

Buena Vista
Bakersfield, California

On a 2,182-acre site in a high-growth area dominated by conventional development, a distinctive urban community of 7,450 residential units has been planned. Traditional town planning principles have been applied to generate a sense of place and shared experience. The development has been laid out as a series of villages that are both self-contained and linked with a three-acre Town Center, which includes a recreation complex with a pool, exercise rooms, arts and crafts studios, etc. Plans also include four elementary schools and one middle school. All parts of the community are connected by an extensive and varied system of parks, totaling 200 acres, many of them featuring bodies of water.

Above left: Town center seen from residential balcony.

Far left: Master plan, with higher density areas shown predominantly yellow, Town Center in middle.

Left: Waterfront structures mixing ground-floor restaurants and retail with upper-floor condos.

Canin Associates Solivita
Poinciana, Florida

"A sense of real community" was a primary goal in planning this age-restricted adult community. The completed first phase of the 3,300-acre, 6,500-unit development comprises a Village Center, which includes two restaurants, a coffee shop, some retail, and a Gathering Center for dances and special events. Further amenities such as a golf course, tennis courts, a grand ballroom, and a craft and fitness center provide centers of community activity that are not dependent on retail. There will be walking and biking paths and a canoe trail along the mile-long community waterway. Intensive landscaping includes thousands of flowering shrubs and numerous canopy trees, many of them relocated from within the site. The streets and architecture of the Village Center appeal through their traditional proportions and planning principles. So effective has this center been as an indication of the entire community that many residents have purchased residences before model homes were even built. Phase III of the community is being completed as this book goes to press.

Facing page, top: Main spine of Village Center.

Facing page, bottom left: Bridge over community waterway.

Facing page, bottom right: Village Center's main entrance.

Above: Characteristic buildings.

Below: Village Center master plan.

Canin Associates

Promenade Town Center
Pasco County, Florida

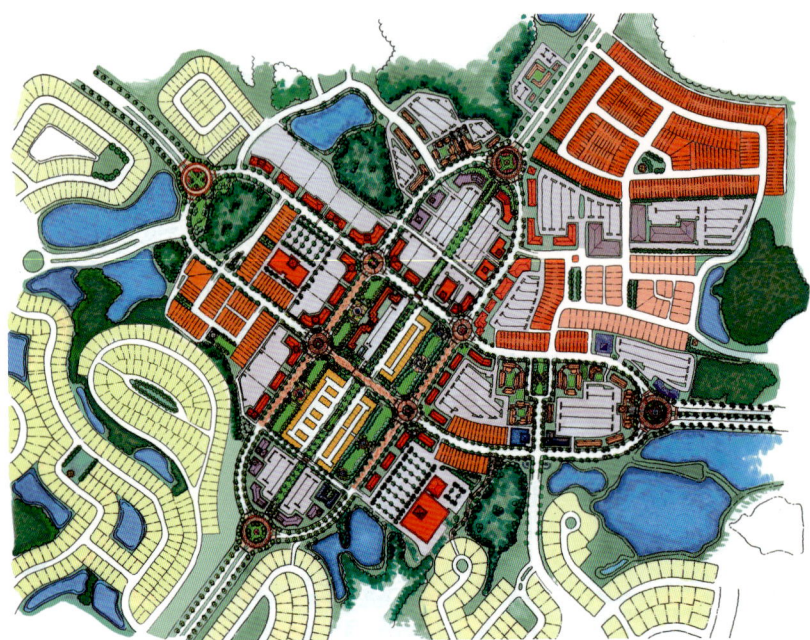

Top: Town center boulevard.
Above: Master plan.

The planning challenge here was to create an appealing mixed-use town center at the intersection of arterial roads that are designed for cars rather than pedestrians. The solution was to lay out two-way couplets to disperse traffic on pairs of one-way streets. Although one-way streets are not usually associated with traditional town planning, in this case they serve the necessary function of reducing street width and traffic flow to a level compatible with a pedestrian-oriented environment. Circles at intersections provide further traffic-calming effects. A pedestrian-friendly tree-lined boulevard – identifiable on the site plan by diagonal green band – runs through the core of the development. The 100-acre town center is planned to include 770 residential units, including townhouses, apartments, and condominiums, plus retail, a grocery store, office space, and a hotel.

Carter & Burgess, Inc.

Corporate Office	Austin	Oakland
777 Main Street	Baltimore	Oklahoma City
Fort Worth, TX 76102	Boston	Orlando
800.494.4082	Columbus	Phoenix
design@c-b.com	Dallas	Raleigh
www.c-b.com	Denver	Sacramento
	Detroit	Salt Lake City
	Fort Lauderdale	San Antonio
	Houston	Santa Ana
	Las Vegas	Tampa
	Little Rock	Washington D.C.
	New York	

Carter & Burgess, Inc.

North Hills
Raleigh, North Carolina

North Hills Mall, one of the South's pioneering enclosed shopping centers, had served affluent Raleigh suburbs for decades, but was being upstaged by newer malls farther out. Developer John Kane saw the opportunity to convert the flagging retail complex into a new "downtown" for those suburbs, strategically located along a major expressway. The $200-million redevelopment started in 2001 with plans to transform the 15-acre North Hills Plaza into an upscale neighborhood shopping center. By 2002 the project had grown to include the 35-acre mall site across the street.

The project includes more than 770,000 square feet of retail and restaurants laid out around a town square and shopping streets with on-street parking. Other components of the project are a 14-screen cinema, a 200-key hotel with banquet facilities, 300,000 square feet of offices, a 300-unit residential building and a 65-unit luxury condominium tower over community-oriented street-level retail. Two levels of underground parking underpin much of the development. In both its planning and architecture, North Hills avoids the homogeneity of the typical lifestyle center, displaying the kind of variety found in a 1920s midtown. Building facades and details are highly varied but limited in color to the neutral tones typically seen in such older cores. The project's pedestrian-scaled streetscape will link its commercial facilities to condominiums on the periphery to form a highly walkable community.

Above left: Internal street, showing fountain and angled parking.

Left: Pedestrian area around central pavilion.

Photography: J. Brough Schamp and Eric Taylor (Right streetscape).

Top: Shopping street with offices above, retail and residential beyond.

Above: Aerial rendering of completed project.

Left: Shops in evening.

Carter & Burgess, Inc.

The Walk
Atlantic City, New Jersey

Left: Aerial panorama.

Below left: Retail shops and fountain.

Bottom left: Street scene showing graphics and landscaping.

Bottom right: Storefronts with bold signage.

Photography: J. Brough Schamp.

Although legalized gambling has spurred construction along Atlantic City's fabled Boardwalk, there have been few attractions to lure visitors away from the casinos and few family-oriented developments. The Walk addressed this problem by converting an eight-block area between the Boardwalk and the new Atlantic City Convention Center into a 320,000-square-foot pedestrian-friendly retail, dining, and entertainment environment. Also linking the city's intermodal transportation center with the Boardwalk, the site had been largely barren and uninviting. The Walk's tenants include shops, restaurants, entertainment facilities, and factory outlets, located in subgroups in an open streetscape enhanced by plazas, fountains, kiosks, and distinctive landscaping, lighting, and environmental graphics. The project was developed by the Cordish Company in collaboration with the city's Casino Reinvestment Development Authority, which made an initial $31-million investment. The two organizations are now researching the viability of adding residential development to the project.

Carter & Burgess, Inc.

Power Plant Live!
Baltimore, Maryland

Above: Overall view showing landmark fountain and identifying sign.

Left: Street scene with bold graphics and generous greenery.

Below left: Restaurant graphics that are lively yet respectful of historic building fronts.

Photography: J. Brough Schamp.

A two-city-block area of twentieth-century row buildings near Baltimore's lively Inner Harbor has been redeveloped as a colorful entertainment district. Adjacent to the established Power Plant retail project, the new district has been dubbed Power Plant Live! and features light shows, a 2,500-seat music hall for nationally known performers, and tenants such as the Improv, Babalu Grill, Maryland Art Place and Ruth's Chris Steak House. Originally, the area's landmark fountain was to be the focal feature, but Carter & Burgess judged that something was needed that would represent the concepts "power plant" and "live." Drawing on the imagery of electricity, the project's design and graphics represent high-voltage energy. The Power Plant Live! sign lights up in sequence, and there are bright marquees and bold neon signs. One entrance is dominated by the image of a workman holding a lightning bolt, which he crashes into an anvil once an hour to set off a light and sound show.

Carter & Burgess, Inc.

Citrus Plaza
Redlands, California

Orange groves, introduced to the area around Redlands in 1873, set off a second California "Gold Rush" and blanketed the area until recent decades, when the mild climate and striking scenery supported a population boom. Located at the intersection of Highways 10 and 30, the 125-acre Citrus Plaza site can draw on a pool of about one million residents. The 50-acre Phase One development, completed in 2004, includes 520,000 square feet, with nationally known retailers along broad landscaped walks. A central fountain and a signature 45-foot tower serve as focal points. Highly visible from the freeways, the project draws design inspiration from the Spanish Colonial architecture of the area. Discreet, abstracted citrus fruit motifs appear in the tops of the openwork metal domes and in lighting elements. The development is planned for expansion to a total of 1.9 million square feet.

Top: Central court and tiled fountain.

Above: Regionally inspired architecture in the landscape.

Left: Ample drives and walkways approaching shops and restaurants.

Photography: Larry Falke (top & left), Richard Leon (above).

Carter & Burgess, Inc.

The Shoppes at Blackstone Valley
Millbury, Massachusetts

For an integrated retail facility that includes big box, junior box, and lifestyle retail, the lifestyle component is laid out in a C-shaped plan that creates a central public space. Building design here is characterized by changes of plane on the relatively continuous storefronts and distinctive roof forms and turrets appearing at key locations. Purposeful over-scaling of the architectural elements and control of signage and lighting lend an air of restraint and dignity to the environment as a whole. The use of fieldstone and brick, along with well-proportioned moldings and eaves, connects the project to regional traditions. The streetscape is generously provided with raised and flush planting beds to humanize the experience of this large-scale complex.

Right: Typical walkway and storefronts.

Below: Buildings around central public space.

Photography: J. Brough Schamp.

Carter & Burgess, Inc.

Walkers Brook Crossing
Jordan's Furniture, Home Depot, IMAX
North Reading, Massachusetts

Above: Jordan's just before opening.

Photography: J. Brough Schamp (above), Anton Grassi (below).

Below: Jordan's Furniture store, IMAX theater, and Home Depot.

Jordan's has become one of the country's highest sales-per-square-foot furniture retailers by making every store a fully themed experience. This 252,000-square-foot facility carries the concept to a new level, with a "Beantown" entertainment area just inside the entrance, an ice cream stand, a full-service diner, and a 500-seat IMAX theater. Playing on Boston's nickname, Beantown exhibits replicas of the city's landmarks composed of over 11 million Jelly Belly jelly beans. While they eat, shoppers can watch a Liquid Fireworks display. The main selling floor resembles a street, its windows displaying complete rooms with coordinated accessories. A glazed volume facing Interstate 95 exhibits the interior activity to the highway. Additionally, this project includes a 165,000 square-foot Home Depot at the lower level.

Chan Krieger & Associates

8 Story Street
5th Floor
Cambridge, MA 02138
617.354.5315
617.354.3252 (Fax)
mhoward@chankrieger.com
www.chankrieger.com

Chan Krieger & Associates Beth Israel Deaconess Medical Center
& Master Plan
Boston, Massachusetts

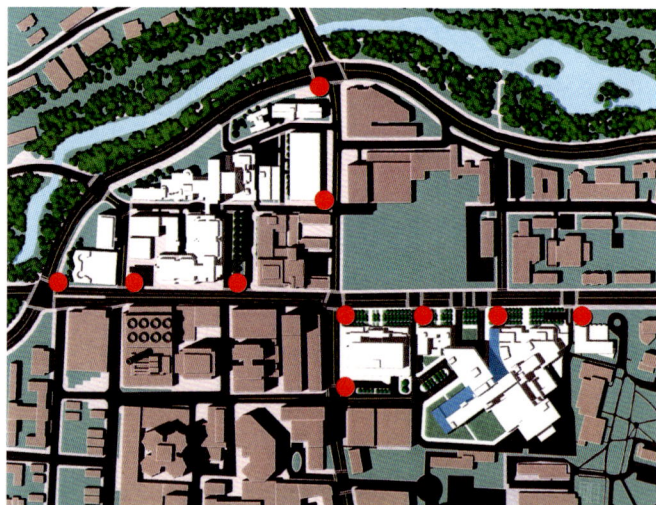

Above: New clinical center, which incorporates former Massachusetts College of Art building.

Above right: Master plan of two combined campuses, with entry points marked.

Right: Landscaped pedestrian link.

Photography: Peter Vanderwarker and Steve Rosenthal.

Above left: New entry court.
Left: Plaza with benches.
Above: Skylighted Clinical Center interior.
Below: Clinical Center atrium.

A $160-million redevelopment program is intended to consolidate the two existing campuses of the former Beth Israel Medical Center and the New England Deaconess Hospital, giving them a single identity. Chan Krieger drew up the master plan for this redevelopment, including design of open spaces, and served as design architect for the shell and core of the Carl J. Shapiro Clinical Center, a new building for the combined institution. The $100-million, 380,000-square-foot clinical center includes medical offices, research facilities, a center for shared technology, and a 500-car underground garage, and it incorporates the former Massachusetts College of Art building. Objectives of the master plan include improving the appearance and efficiency of the unified campus, introducing design features to define its entry portals, and distinguishing it within the larger context of the Longwood medical facilities area and the city.

Chan Krieger & Associates

City Hall Plaza Community Arcade and Government Center Master Plan
Boston, Massachusetts

Left: Arcade, showing cable-supported canopies.

Right: Overall view with Cambridge Street at left, JFK Federal Office Building beyond arcade.

Below: Seating platform built over granite plaza steps.

Photography: Chan Krieger & Associates, Patrick Whittemore/Boston Herald.

The arcade is the first phase of a larger City Hall Plaza redesign and renovation effort master planned by the firm. The arcade is designed to better define the curved Cambridge Street side of the plaza without closing it off. It will serve as a spine between two proposed projects: a new transit station at one end and the JFK tower garden at the other. Included in the arcade project are

Left: Arcade as community open space.

Below: Arcade on market day.

three seating platforms, extending beyond the existing plaza steps to serve as public seating areas, stages for small performances and speakers, or stalls for periodical farmers' markets. An elaborate lighting scheme includes LED light tubes mounted on the arcade columns, which can be programmed to change colors. The plaza master plan entailed a four-year process of studies and public approvals. It seeks to improve the "Walk-to-the-Sea" from Faneuil Hall to the harbor, to create small-scaled gathering and lingering spaces, to enhance the amphitheater-like qualities of the plaza for public activities, and to reintroduce Hanover Street to make the plaza a more effective part of downtown movement patterns.

Chan Krieger & Associates

Three Rivers Park
Pittsburgh, Pennsylvania

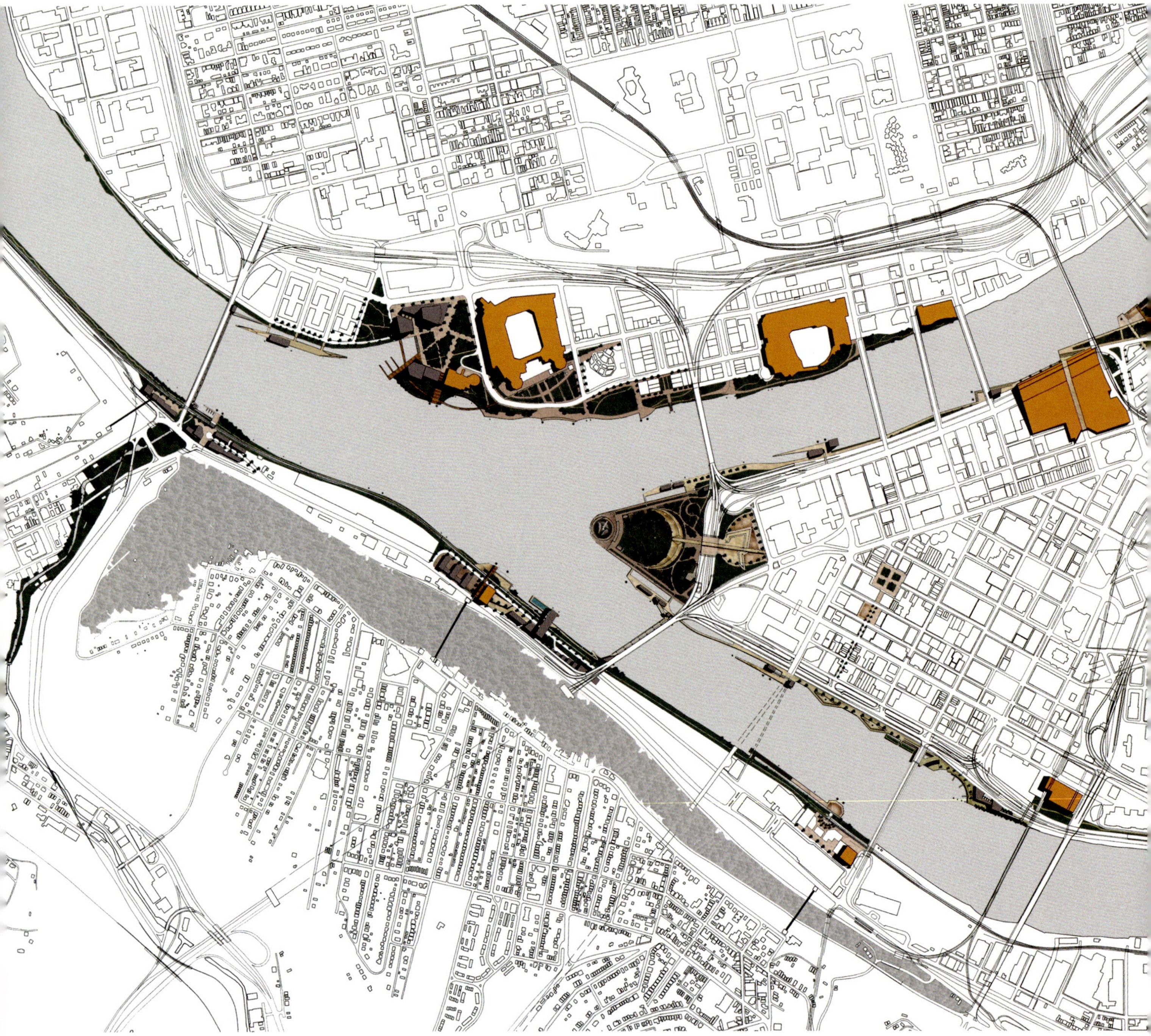

Top: Proposed north shore development.

Above: Master plan showing proposed public and private development.

Left: Bessemer Court Pedestrian Bridge on south shore.

The RiverLife Task Force chose the firm to create a vision plan for energizing and directing development along Pittsburgh's three riverfronts. With two new stadiums, a new convention center, an extension of a light rail system, and several corporate buildings in design or under construction, a primary goal of this effort was to ensure that new private development and public investment will be carried out within a sustainable framework and to world-class design standards. New housing, water transit, and overall balance of public amenities and private investments were reviewed through an extensive public process. The resulting plan refocuses the Pittsburgh community on its rivers and proposes a great urban river park. With the backing of the city's major foundations, institutions, and corporations, the task force is charged with directing public and private investments totaling $10 billion over the next 10 years. The firm also designed the Bessemer Court Pedestrian Bridge, located at Station Square, a 52-acre entertainment development across the river from downtown Pittsburgh's Golden Triangle. The project includes $3 million in public funding for improvements, including a pedestrian bridge spanning railroad tracks for access to the riverfront, which features a public marina, a floating boardwalk, and an entertainment venue called The Landing. Designed to provide views of Bessemer Court's fountains, the river, and the city, the bridge features contemporary industrial materials that also reflect the industrial heritage of the site.

Top to bottom: Tip of Golden Triangle, showing stadiums on north shore; Golden Triangle with lighted bridges; two aerial views of proposals for confluence of rivers.

Photography: Chan Krieger & Associates.

Chan Krieger & Associates

Fort Washington Way Highway Reconfiguration
Cincinnati, Ohio

Top left: Bridge over highway showing new design elements.
Above left: Benches and tree plantings along walkway.
Left: Mast and barrier on bridge.
Above: Bridge details.
Photography: Chan Krieger & Associates, Parsons Brinckerhoff, Hargreaves Associates.

This $160-million project is associated with the reconstruction of Fort Washington Way, part of an interstate highway along the edge of downtown Cincinnati. Covering 30 acres of highway and adjoining service roads, the project includes: five new bridges — two of them designed as new city "gateways"; over a mile of roadway walls providing visual relief for motorists and pedestrians; custom-designed light fixtures and railings integrated into the overall aesthetic yet complying with stringent federal highway standards. Linear parks along the surface boulevards flanking the highway, developed in collaboration with Hargreaves Associates, include: tree plantings, walls for sitting, special paving materials, protective barriers, and signage/banner standards.

Charlan•Brock & Associates, Inc.

2600 Maitland Center Parkway
Suite 260
Maitland, FL 32751
407.660.8900
407.875.9948 (Fax)
Butch@cbaarchitects.com
www.cbaarchitects.com

Charlan·Brock & Associates, Inc.

Williams Walk at Bartram Park
Jacksonville, Florida

Even though mixed-use developments are today's exciting housing types, there remains a strong market for stand-alone garden apartment communities. For developers, they offer the ability to respond quickly to changing local economic conditions. This 380-unit Williams Walk community represents this type of development and market situation. Residential buildings have been carefully sited to maintain the feeling of several villages. Views of the lake have been left unencumbered and parking has been sited to minimize its visual impact. The architecture of the project reflects modern design philosophies and historical influences, drawing inspiration from the nearby historic town of St. Augustine and the travels of the early botanist, William Bartram. The development's buildings are nestled among large existing oaks and pines. Outdoor spaces are designed as roofless rooms, furnished with elements that evoke history. A variety of cedar gateways are embellished with hand-painted botanical views. The Village Meeting Hall, or Clubhouse, evokes an earlier era of subtropical leisure. Without leaving the property, a resident can lounge in a hammock or club chair in a shady arbor, dine al fresco around an outdoor fireplace, indulge in a spa getaway, or swim laps in the community swimming pool.

Above right: Site plan, with entrance at right, clubhouse nearby on lake.
Above left: Cedar gateway featuring botanical painting.
Left: Residential courtyard
Right: Clubhouse loggia overlooking pool.

Top: Community pool and clubhouse.

Above: Portion of pool area.

Right: Typical residential buildings and landscaping.

Photography: Beverly Brosius.

Charlan·Brock & Associates, Inc.

Aqua Condominiums
Panama City Beach, Florida

Left: Ground floor plan of tower and garage.
Below left: Beachfront view.
Below: Entrance front.
Renderings: Genesis Studios.

An unusually long, thin oceanfront site conditioned the design of this high-rise resort condominium. The 5.37-acre property was so restrictive that parking had to be located in a parking garage on the opposite side of a busy highway, with a safe, secure connection via a 200-foot-long bridge. The narrow configuration of the site also left inadequate space for the desired beachfront pool deck. This limitation was converted to an asset by carving some of the pool environment into the base of the building. Water was made an element of the Aqua theme with a series of fountains, pools, and runnels leading from the main entrance through to the pool deck. Other identifying features of the design are the broad trellises mounted on the rooftop, providing a play of light and shadow over the building surfaces. Another design focus was the use of salt-tolerant materials such as the split-face limestone cladding on the structure's lower floors. Budgeted at $55 million, the project is scheduled for completion in 2006.

Charlan·Brock & Associates, Inc.

Uptown Maitland West
Maitland, Florida

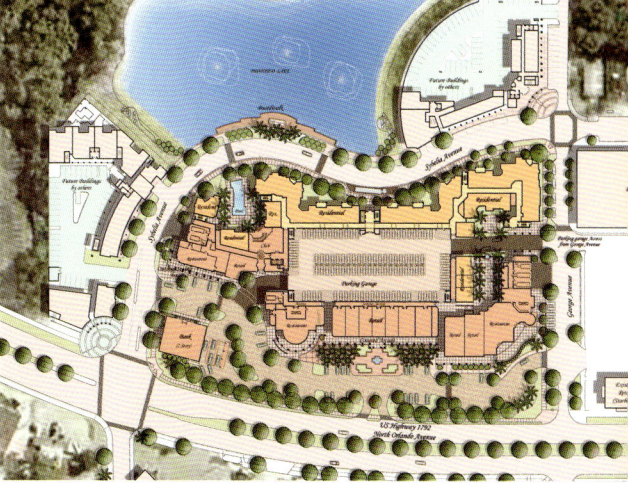

Above: Site plan.

Left: Perspective at major intersection, lower right on site plan.

Renderings: Genesis Studios.

Above: Partial elevation on George Avenue, right on site plan.

Below: Elevation on North Orlando Avenue, bottom on site plan.

Uptown Maitland West is an important step in revitalizing the downtown of Maitland, a suburb of Orlando. Located on 6.13 acres, this private development in the heart of the city is designed to contribute to a vibrant Town Center. It includes 300 condominium homes, 34,000 square feet of retail, 3 restaurants, and a bank. The development has been conceived as one where many residents will also work in the community's shops and restaurants. Planning has been fine-tuned to ensure that the pedestrian takes precedence over vehicular traffic. Estimated to cost $65 million and slated for completion in 2007, the project is envisioned as inspiring continued redevelopment of Maitland's core.

Charlan·Brock & Associates, Inc.

Rarity Pointe Lodge and Spa
Knoxville, Tennessee

Left: Aerial perspective of development.

Below left: Site plan.

Bottom right: Elevation of Lodge interior.

Bottom left: Inspiration for niches on interior elevation.

Renderings: Genesis Studios – Peter Chanakul.

A hilltop village overlooking Lake Tellico, just outside Knoxville, will evoke the qualities of English country estate of around 1900. Working with an interdisciplinary team, the architects have organized almost 200 condominium units on the five-acre site. Most of the units will be in two five-story structures flanking the central Lodge. The Lodge itself will include 13 condominiums above common facilities that will be open to the wider community: a 5,500-square-foot health club and spa, open to a wider public, a 20,000-square-foot conference facility, and a 7,000-square-foot first-class restaurant conceived in the spirit of the Biltmore estate in Asheville, North Carolina. The Lodge complex will include indoor and outdoor pools, and there will be extensive boating facilities along the lake shore.

Charlan·Brock & Associates, Inc.

The Flats at Rosemary Beach
Rosemary Beach, Florida

This 60-unit resort condominium development had to meet the stringent design guidelines of Rosemary Beach, which call for buildings recalling Florida and West Indian traditions. The detailing required under guidelines can be expensive, but the architects were able to interpret them here within the developer's target sales price range. One strategy was to build simple, repetitive volumes. Most of the buildings contain two mirror-image units per floor, stacked three high, each offering three exposures. An appropriately tropical exterior image, with generous porches, has been created using straightforward components. All of the units overlook a central park, with small-scaled parking areas on their opposite fronts.

Above: Site plan.

Right: Typical exterior treatment, with turret on end unit.

Left: Walled ground-floor terraces and bracketed upper-floor porches.

Below: Shared central park.

Below right: Subtly varied row.

Photography: Charlan·Brock & Associates, Inc.

Charlan·Brock & Associates, Inc.

Cheval Apartments on Old Katy Road
Houston, Texas

Left: Site plan, with fault line toward lower left.

Below: Main entrance front, with arched entrances to motor court.

Bottom: Overall front view.

Renderings: Genesis Studios.

As land values escalate, a client who had previously built simply-designed garden apartments took on a more demanding challenge, an old coffee mill on a 10.5-acre site that has a geological fault line running through it. The mill will be replaced by a 387-unit urban apartment development. The project will project the image of a single low-rise structure, wrapping discreetly around a central parking structure. Given the number of units, many will necessarily be oriented inward, where inviting courtyards have been created. The extensive corridors have been treated as pedestrian streets, with intersections designed to promote socializing and some routes passing through the club facilities. A motor court, just inside the entry arches, establishes a distinctive first impression and helps with orientation.

88

CBT/Childs Bertman Tseckares Inc.

110 Canal Street
Boston, MA 02114
617.262.4354
617.236.0378 (Fax)
www.cbtarchitects.com

CBT/Childs Bertman Tseckares Inc.

Columbus Center
Boston, Massachusetts

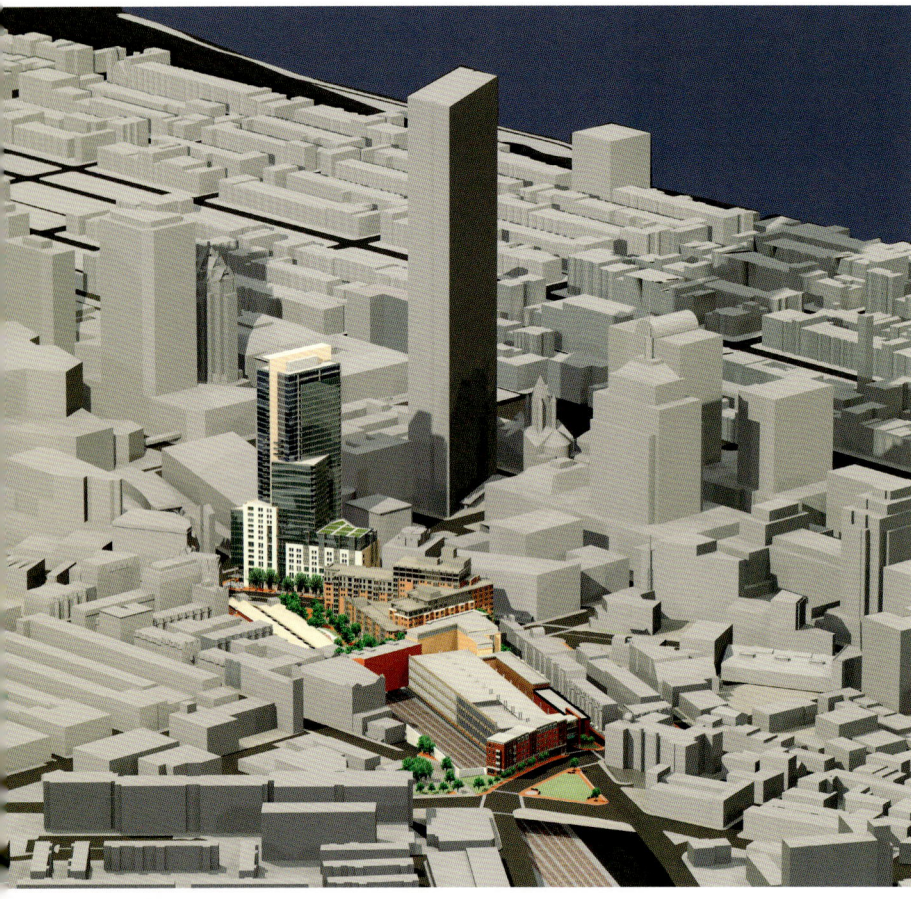

Above: Project in city context.
Above: right: Mixed-use tower.
Left: Residential street.
Below left: Turnpike topped by residential buildings surrounding garage.

This 1.4 million-square-foot mixed-use development is proposed for three parcels — equivalent to three city blocks — over the Massachusetts Turnpike. It will knit together the historic South End and Back Bay neighborhoods. The parcels closest to the existing commercial development will include two high-rises with ground-floor retail, hotel floors above the retail, and several stories of luxury condominiums at the top. The third parcel will contain a 650-space parking garage surrounded by townhouses and retail. The project is designed so that it "looks like city streets and not like a mega project." Columbus Center buildings have individual, rather than group, identity — the high-rise is light and airy, and the lower buildings are detailed in masonry like the existing context. Bridging over the turnpike right of way posed major engineering and design challenges. In order to avoid elevated first floors, the huge support structures spanning the highway have to be concealed inside the buildings rather than under them.

CBT/Childs Bertman Tseckares Inc.

North Point
Cambridge, Boston, Somerville, Massachusetts

Top: Bird's-eye view of entire project.

Above: Loft-style residences.

Left: Views of park and mixed-use buildings.

North Point is envisioned as a unique 21st century smart-growth, transit-oriented neighborhood that will transform 45 acres of under-utilized industrial land in the municipalities of Cambridge, Boston, and Somerville, MA, into a vibrant and integrated mixed-use community. The development consists of over 2.2 million-square-feet of commercial space, 2,700 residences and substantial retail to support 24-hour activities of residents, workers and visitors. The master plan also encourages a variety of built-form and architectural expression that is reflective of a true urban setting. Once complete, this 15- to 20-year project will have created 20 new blocks, a new Square for a state-of-the-art subway station, a multi-use bike and recreational trail, nearly one mile of new roadway and utility infrastructure, and a wide-range of public amenities. At the heart of North Point is the Central Park, 6.3-acres of green space designed to serve those that live and work in the neighborhood, as well as to attract visitors from the surrounding area.

CBT/Childs Bertman Tseckares Inc.

The Residences at Kendall Square
Cambridge, Massachusetts

Left: Overall view of tower.

Below: Street level, with shops and residential entrance.

The Residences at Kendall Square will be one of the first projects built in a dynamic new 10-acre, mixed-use community in Cambridge, MA, conceived as a hub for biotechnology. Additional elements of the project are offices and laboratories for biotech companies, and extensive entertainment and recreation facilities, such as restaurants, outdoor parks, and a small kayak facility on a canal that feeds into the Charles River. CBT has designed the tallest building in the new development, a 332,000-square-foot residential tower that will include 330 rental apartments ranging from studios to two-bedroom units for employees and students of nearby universities and research centers, as well as 10,000-square-feet of retail space and a 15,000-square-foot health spa. While the top of the building animates the Cambridge skyline, the one-story base juts out and away from the tower and connects into a lively pedestrian network of open spaces, promenades, specialty shops, and restaurants at ground level.

CBT/Childs Bertman Tseckares Inc.

Rollins Square
Boston, Massachusetts

With a strong commitment to provide residential opportunities for families of all incomes, the Planning Office for Urban Affairs/Archdiocese of Boston developed Rollins Square as a model for a mixed-income community that combines market-rate, moderately-priced, and low-income housing in a high-quality condominium complex. CBT designed this award-winning project as a grouping of six-story elevator buildings and four-story townhouses to foster a sense of community for the residents. In addition to 184 residential units that vary in size from one-bedroom apartments to three-bedroom duplexes, the 376,000-square-foot project also includes a 200-space below-grade parking facility in an area where parking is scarce. Uniting the Victorian architectural traditions of the historic district with the industrial character of nearby warehouses that surround the site, Rollins Square harmonizes with the existing cityscape without overwhelming it. At the center of the project, the buildings wrap around a quaint exterior courtyard that captures the spirit of the surrounding district and gives the residents a shared sense of community, security and place. The complex also maintains a pedestrian orientation to the neighborhood by providing new ground-level retail and office space, and seating walls that serve as social gathering places for residents and members of the larger community.

Top: Complex seen along existing street.

Above: Aerial view.

Right: Buildings around central courtyard.

Photography: Robert Benson; Mark Flannery (aerial).

CBT/Childs Bertman Tseckares Inc.

The Prudential Center Redevelopment
Boston, Massachusetts

CBT is leading the effort on a 15-year, $1 billion expansion and redesign of the Prudential Center. Upon completion, the 27-acre site will feature 1.8 million square feet of new office, residential, and retail space that will give this 40-year old Boston landmark a new identity.

111 Huntington Avenue is a 21st-century icon that bolsters the Prudential Center's identity as a world-class commercial district in Boston. The 36-story office tower is the cornerstone of CBT's revised master plan for the southern portion of the complex.

The Winter Garden serves as the Prudential Center's dramatic new entrance for Huntington Avenue. The Garden is lined with retail shops, pedestrian connections and seating areas under a glazed façade and roof stretching 50 feet wide, 35 feet high and over 400 feet long.

The Belvedere is a 131,000-square-foot 11-story luxury condominium building that offers convenient 24-hour access to the Center's urban amenities and cultural activities in addition to a variety of other services that support a city lifestyle.

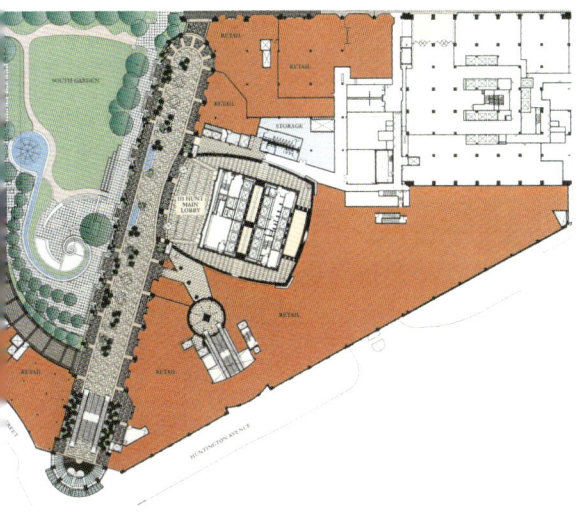

Facing page: 111 Huntington Avenue, Boston's newest addition to it's skyline.

Left: Partial master plan, showing Winter Garden, landscaped court, and lower floors of Belvedere and 111 Huntington Avenue.

Below: 111 Huntington Avenue in context, with original Prudential tower in background.

Right: The Belvedere.

Right middle and bottom: Exterior and interior views of Shaw's Supermarket.

Above: Huntington Avenue entrance.

Above right: A portion of the 400-foot-long Winter Garden.

Below: The Mandarin Oriental (rendering by Neoscape).

Photography: Jonathan Hillyer/Esto, 111 Huntington Avenue and The Belvedere; Edward Jacoby, Shaw's Supermarket.

The Mandarin Oriental will be a 450,000-square-foot building that will include a 150-room Mandarin Hotel, 50 individually planned condominiums, 35 rental apartments, street-level retail space, and guest amenities to define the highest standard of luxury in Boston. The Mandarin Oriental-Boston will be a microcosm of the 24-hour urban environment that has come to distinguish the Prudential Center as a place to work and live.

Shaw's Supermarket provides a vital urban amenity for the surrounding residential area. Situated in a prominent and accessible location within the Prudential Center complex, the 52,000-square-foot facility combines a range of amenities with the size of a superstore yet offers the convenience and breadth of product typically found in the suburbs.

Costas Kondylis and Partners LLP

31 West 27 Street
New York, NY 10001
212.725.4655
212.725.3441 (Fax)
info@kondylis.com
www.kondylis.com

Costas Kondylis and Partners LLP

Costas Kondylis and Partners LLP Trump World Tower
845 United Nations Plaza
New York, New York

A neighbor of the United Nations Headquarters in Midtown Manhattan, this 860-foot-tall residential tower rises as a sleek rectangular form, without setbacks or ornamentation. Designed to appear as if sheathed in a single sheet of glass, the building was inspired by such nearby International Style landmarks as the Seagram Building and the U.N. Secretariat. Its curtain wall is composed of deep-bronze-colored glass, with a structural silicone glazing system, creating a subtle, incised grid that gives the walls scale and texture. On the interior, glazing extends from the ceilings — ranging from 10 to 16 feet in height — down to 20 inches above the floors, offering residents panoramic views of the Manhattan cityscape, the

East River, and distant landscapes. The uninterrupted verticals of the tower's form and an advanced structural system allow optimum layouts of the 370 apartments, without the obstructions of massive piers or setbacks. Street-level amenities include a 30-foot-high lobby, a generous porte-cochere, and a private garden. There is also a 72-car garage, an 11,000-square-foot health club with a 60-foot lap pool and a ground-floor luxury restaurant.

Facing page: Tower rising above nearby buildings.

Above: Building as high point on East Midtown skyline, U.N. Headquarters to left.

Left: Upper floors of shaft, showing curtain wall details.

Photography: Jeff Goldberg©. ESTO (facing page).

Costas Kondylis and Partners LLP The Heritage
New York, New York

Above: Trump Place along Hudson.

Below and facing page: The Heritage, seen from Riverside Park with other Trump Place buildings beyond.

Photography: Chuck Choi.

Occupying a unique site, where the Manhattan street grid meets the Trump Place development in a sweeping curve, the Heritage is designed as the unmistakable cornerstone of the towered row. Rather than designing the building by fitting units into its curved volume and the square tower that rises from it, the architects took an "inside-out" approach in which the optimum apartment layouts were arrived at first. Features of the 174 units include foyer galleries, studies adjoining bedrooms, and luxurious bathrooms with "his" and "hers" dressing rooms. To make the most of dramatic views, window assemblies seven feet high and 16 feet wide were adopted. Recessing balconies into the façade creates shadowed variations in the façade pattern and provides occupants with outdoor spaces protected from wind and rain. Two full-floor units on the 30th and 31st floors feature loggias 30 feet long and 8 feet deep. Amenities of the Heritage, consistent with those of other Trump Place buildings, include meeting rooms, a children's play room, storage rooms, a 199-car garage, and a second-floor health club, with two indoor pools offering stunning views of Riverside Park and the Hudson River.

Costas Kondylis and Partners LLP Morton Square
600 Washington St.
New York, New York

To build a full-block residential complex in an area traditionally zoned for manufacturing, the developers first went through a two-year process to get a special permit. An agreed-upon zoning envelope for the site allowed for three contiguous buildings of different types. A 14-story L-shaped tower contains 129 condominiums. A six-story block contains a row of six three-story townhouses, topped by an additional three stories with 12 loft units. A seven-story rental building houses 136 apartments. The three structures enclose a lushly landscaped shared garden over a two-level 133-car garage. A major design challenge was to link the diverse buildings with common architectural elements, principally horizontal bands of limestone-colored precast concrete, uniform metal panels and window frames, and transom glazing at the tops of windows. In the tower building, floor-to-ceiling glazing, curved corners, and glazed projections visibly express the goal of maximizing Hudson River views. The townhouses have bowed or projecting fronts, with private stoops, recalling the city's typical rowhouses. The fronts of the loft units above are compatible but clearly in the tradition of industrial lofts. The rental building is simpler in detail, while sharing the rusticated base and piers of the same precast material.

Facing page: View of townhouses with lofts above.

Above: L-shaped condominium tower, apartment block at right.

Right: View of rental building.

Photography: Chuck Choi.

Costas Kondylis and Partners LLP The Grand Tier
1930 Broadway
New York, New York

Sited prominently on Broadway opposite Lincoln Center, the Grand Tier contains 230 luxury rental apartments, street-oriented retail, a health club, and a parking garage. Total floor area is 448,000 square feet. The podium-and-tower massing was determined in part by a special zoning permit, required for its Lincoln Square area site. While the podium maintains the street line along Broadway to a prescribed height of 85 feet, the tower is set well back and configured as a cluster of bays, exploiting the extraordinary views east toward Central Park and west toward Lincoln Center and the Hudson. Exterior limestone and limestone-colored masonry were chosen to echo the hue of the performing arts center's travertine cladding. Because of the depth of apartments in the podium portion, they were designed as loft-type units, with high ceilings and broad windows facing Lincoln Center. Several of these units are duplexes, as are those in the top two stories of the tower. Amenities available to tenants include a swimming pool, a lounge and an outdoor roof deck.

Left: View across Broadway toward Grand Tier.
Photography: Chuck Choi.

Cunningham + Quill Architects PLLC

1054 31st Street NW
Suite 315
Washington, DC 20007
202.337.0090
202.337.0092 (Fax)
marketing@cunninghamquill.com
www.cunninghamquill.com

Cunningham + Quill Architects PLLC

Cunningham + Quill Architects PLLC

The Mather Building
Washington, DC

Left: Restored terra cotta detail and new canopy.

Below: Restored street façade.

Right: Duplex apartment.

Below right: Penthouse details and view of Washington Monument beyond.

Photography: Maxwell MacKenzie.

This project set a precedent as Washington's first conversion of a downtown office building into housing. Constructed in 1917, the structure was occupied by the University of the District of Columbia from 1967 to 1989. The adaptive reuse completed in 2003 provides for arts spaces on the street level and residential uses above. About one quarter of the residential spaces are live/work units suitable for artists, the remainder market-rate apartments for contemporary living. A penthouse, added as an approved exception to the DC height restriction, offers spectacular views of the National Mall. As a contributing structure in the Downtown Historic District, the building required – and unanimously won – approvals of the DC Historic Preservation Review Board, the Board of Zoning Adjustment, and the Area Neighborhood Commission. The terra cotta of the notable Gothic Revival façade was painstakingly restored on the basis of historical research.

Cunningham + Quill Architects PLLC

Huntfield Master Plan
Charles Town, West Virginia

Left: Master plan.

Below: About half of 995-acre development, with rail line and possible station in foreground.

Right: Historic core of Charles Town.

Bottom left: Completed houses.

Bottom right: Huntfield Monument neighborhood.

Photography: Christopher Morrison, Greenvest LC.

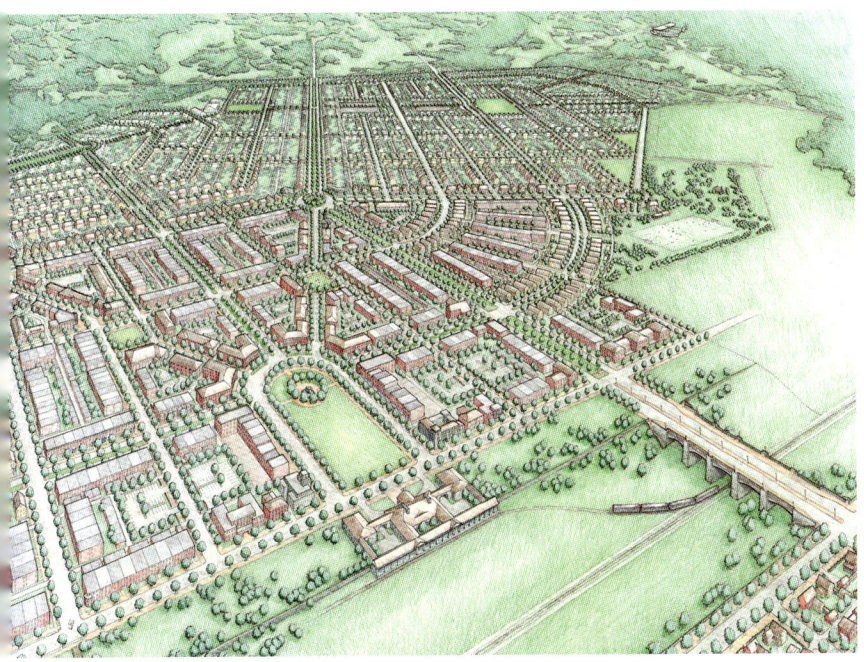

Following county rejection of the owner's previous plan for the 995-acre site, CQA was commissioned to lead a new master plan team. The site is about 1¼ miles from the center of Charles Town, immediately adjacent to Claymont, an impressive c. 1820 house built by George Washington's nephew. The plan reflects a study of communities in the region and embodies the principles of smart growth: compact design, mix of uses, and clear links to the existing community. Its neighborhoods are based on ¼-mile (5-minute) walking distances, with parks, retail, and civic facilities as neighborhood centers. Views of Huntfield from Claymont and other key vantage points will show house fronts rather than backyards, and high points will be occupied by public spaces rather than houses. The firm is exploring the possibility of extending an existing commuter rail along a track that traverses the site, with transit-oriented development around an on-site station. The first neighborhood is now completed and selling well.

Cunningham + Quill Architects PLLC

Caton's Walk
Washington, DC

Located in Georgetown along the historic C&O Canal National Park, this building was constructed in 1929 as a commercial automobile garage. Although the street front is of brownstone, the walls exposed to the canal are of fieldstone with a brick upper story. Caton's Walk extends the redevelopment of old commercial structures that began decades ago with Canal Square. The immediate area is now the site of several interior furnishings showrooms. Approval for building modifications was obtained from the Old Georgetown Board and the U.S. Commission of Fine Arts, with support from the National Park Service and the local community. CQA oversaw the historic preservation of the structure and its conversion to retail, office, and residential uses, with two luxury units in a third-floor addition. Steel-framed windows and the steel-and-glass penthouse exterior were approved as in character with the building and its industrial context.

Above: Caton's Walk from canal.

Left: Stone wall and penthouse addition.

Below: Steel details on old masonry wall.

Right: Terrace outside set-back penthouse.

Photography: Christopher Morrison.

Cunningham + Quill Architects PLLC

The Alta
Washington, DC

Above: Exterior detail with loft-like residential unit.
Left: Typical unit plan
Below left: Rooftop terraces.
Below: 14th Street elevation.
Renderings: Interface

Currently under construction in the downtown area near Thomas Circle, the Alta is a 13-story concrete-framed high-rise building with retail uses on the ground floor and 126 residential units above. Introducing residential uses to a formerly all-commercial district, the project is designed to appeal to young professionals who are interested in an urban setting and loft-like spatial qualities. The typical unit is divided by a linear core that includes all kitchen appliances and other mechanical requirements. Amenities include a community room and landscaped terraces at the second floor shared by all occupants, with private terraces at the top offering views of the National Mall. The exterior palette of stone, precast, metal, and glass is detailed to suggest urbanity and luxury. The design includes a planted roof, and the project team is working toward LEED certification for its environmental qualities. If approved it would be the first such recognition for a residential building in the District of Columbia.

Cunningham + Quill Architects PLLC

National Cathedral School
Washington, DC

In a competition design for this well established girls' school, CQA proposed a phased process of demolition, remodeling, and new construction that reconcile the school's motley physical plant with the larger National Cathedral campus. The firm's design organized the school's upper, middle, and lower divisions around a courtyard open on its fourth side toward views of the hilltop cathedral and secluded from busy Wisconsin Avenue. Each of the three divisions would have had its own distinct entry. Proximity to a new theater was intended to facilitate use of the courtyard for performances. Arched openings, pitched roofs, and prominent clusters of chimneys recall historic precedents. The masonry walls were to be graduated from largely brick on the existing exterior of the lower school to brick with large areas of limestone on the new upper school wing — linking it visually with nearby Hearst Hall, which also houses much of the upper school program. Generous areas of glass brighten the interiors, at once expressing the buildings' modernity and its affinity with the cathedral's High Gothic.

Right: Cathedral and neighboring existing building.

Below left: Area plan of cathedral and outbuildings.

Below center: Site plan of school.

Below right: Perspective of school wings around new entry court.

Photography: David Bagnoli.

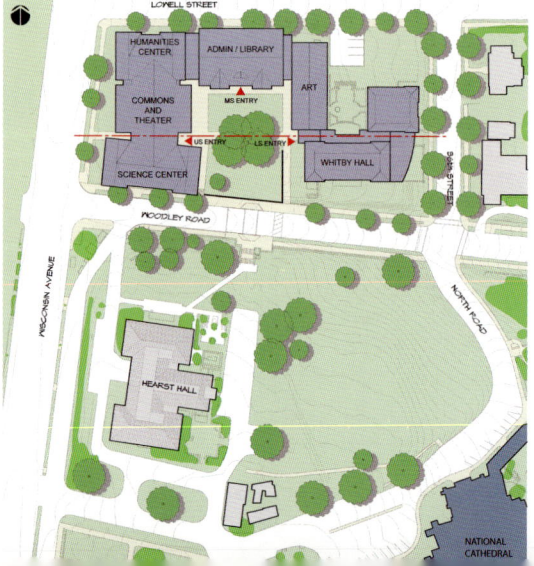

Cunningham + Quill Architects PLLC

Park Hill Condominiums
Washington, DC

This 29-unit condominium on Connecticut Avenue reconciles two very different architectural contexts: a street wall of masonry apartment houses, including some of Washington's official "best addresses", and a number of freestanding Modernist buildings across the avenue, including the Intelsat headquarters. The new building is composed of two five-story masonry "bookends" and a central volume with a metal and glass façade rising eight stories from a shallow entry court. The different portions of the building are unified by devices such as the proportions of the tall windows and the projecting metal balconies on the masonry blocks. Because the building is close to a Metro transit station, it was allowed to rise from the prevailing 60 feet to 90 feet over part of the site, affording the upper-floor units views of Washington's major landmarks and the National Mall. Modernist interiors with fluid spaces introduce a new paradigm to this traditionally "best address" residential area, anticipating the market for loft-like units that would soon follow.

Above: Duplex units on upper floors.

Below: Typical condominium interior.

Above right: Building front in context.

Right: Glass-canopied main entrance.

Photography: Daniel Cunningham.

Cunningham + Quill Architects PLLC

Fortnightly Neighborhood Master Plan and Herndon Senior Center
Herndon, Virginia

Top left: Fortnightly Boulevard leading toward Senior Center.

Top right: Aerial view with Senior Center in foreground.

Center: Elevation of Senior Center.

Left: Neighborhood plan.

Above: New Senior Center by CQA.

Photography: Scott Matties.

Renderings: Luc Herbots.

Working with the Fairfax County Department of Housing and Community Development, CQA planned a mixed-use neighborhood adjacent to the historic downtown and new municipal center of Herndon. Revitalizing 11 acres of mostly vacant and blighted industrial properties, the plan provides an urban village connection between the downtown and a newly constructed senior housing development. The plan proposes a variety of single-family town-homes, multi-unit housing and office/retail buildings along tree-lined streets, with alley access to garages. Rezoning of the area for mixed use involved garnering the support of community residents and leaders. To ensure high quality development, the design team created design guidelines covering a range of issues from building massing to fenestration and street furniture. Some prototype housing is now completed. A new senior community center by CQA has been completed and includes activity and meeting rooms, a commercial dining facility and double height multi-purpose room, and serves as a physical functional link between the senior development and the community as a whole.

Dahlin Group
Architecture Planning

5865 Owens Drive
Pleasanton, CA 94588
925.837.8286
925.837.2543 (Fax)

ddahlin@dahlingroup.com
www.dahlingroup.com

415 South Cedros Avenue, Suite 200
Solana Beach, CA 92075
858.350.0544
858.350.0540 (Fax)

Digital Imaging Studio
101 Townsend Street, Suite 209
San Francisco, CA 94107
415.538.0933
415.512.1313 (Fax)

18818 Teller Avenue, Suite 260
Irvine, CA 92612
949.250.4680
949.250.8002 (Fax)

Suite 502, Tower C
Heqiao Mansion
Jua #8, Guang Hua Road
Chaoyang District
Beijing 100026
China
011.86.10.6581.0466
011.86.10.6581.0470 (Fax)

Dahlin Group

Dahlin Group

Coyote Valley
San José, California

Coyote Valley is envisioned to be a high-density pedestrian and transit oriented urban community with a population of approximately 75,000 residents in the City of San José. A restored four-mile long Fisher Creek and a 50-plus-acre focal lake, park, a 1.8-mile canal/park system, a fixed guideway transit system and a multi-functional Parkway system are planned for the 7,000-acre site, of which 3,500 acres are planned for urban development. Originally conceived as a way of dealing with storm water runoff and flood control, the lake, canal, and creek have become defining features of the community. The heart of the Coyote Valley is envisioned as a mixed-use urban core focused on the lake that will include residential, office, commercial uses, plus plazas and parks. Circulation will follow a merge and loop parkway system linking grids of low-volume, pedestrian-friendly streets. The entire community will be within 1,500 feet of the fixed-guideway transit system, initially using open style vehicles, linking the community to the regional public transportation system. Housing will range from high-density single-family detached units (10 dwelling units per acre) to 20-story towers (100 dwelling units per acre), with greater than 72% of the units at densities of 22 per acre or greater. Coyote Valley will also include a minimum 20 percent of the units as deed-restricted for affordable housing. The full build-out, over approximately 30 years, will include a minimum of 25,000 dwelling units and 50,000 industry-driving jobs with a total of about 60 million square feet of construction. Also included in the plan will be nine elementary schools, two middle schools and one collegiate style high school containing several smaller campuses.

Right: Master plan.

Opposite center: Rendering of full community, with creek leading to lake at center.

Below: High-density mixed use surrounding lake.

Bottom: Proposed street elevations.

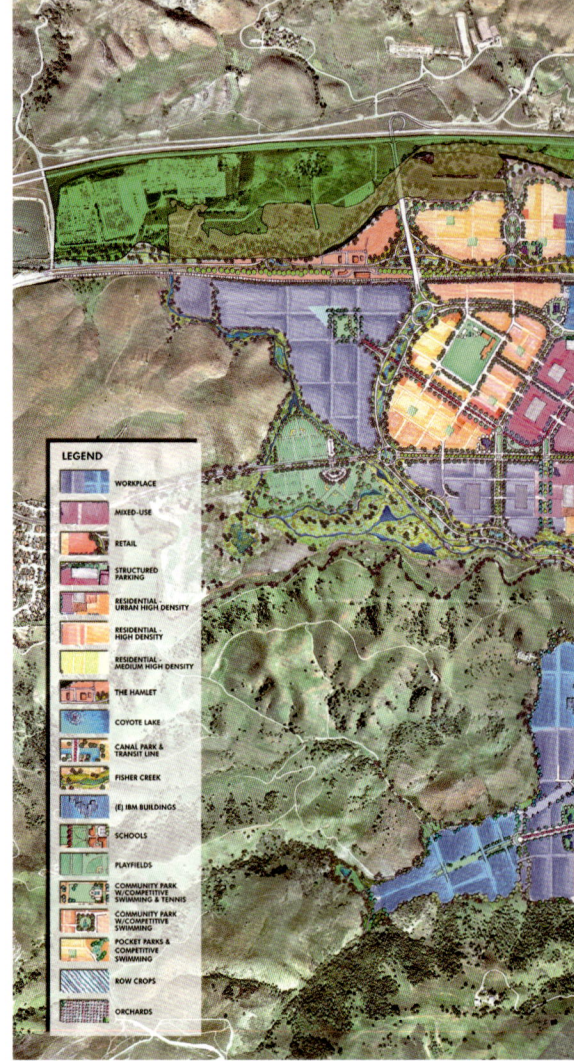

Dahlin Group

Luxe Hills International Golf Community
Cheng Du, Sichuan Province, China

Left: Two-level driving range at public golf teaching center.

Below: Detached villas varying in configuration.

Photography: Wide Horizon Real Estate Development Company.

Above: Community center in the Wan An style, using local river stones, stucco, teak, and tile roofs.

Below: Overall master plan provides a community skeleton of networked and focal amenities. The circulation has been designed to capture powerful vistas of these amenities, key focal landmarks and unique site features.

Facing page, bottom: Community gatehouse.

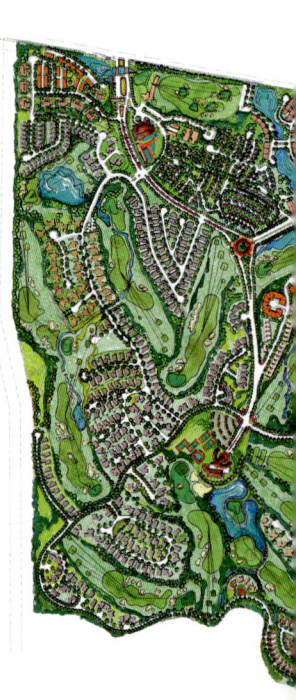

An emerging generation of affluent, well-traveled people in China has generated a demand for living environments of international quality. To meet this demand in the Sichuan Province, Luxe Hills' master plan includes a championship 18-hole golf course with clubhouse and the widest possible range of residential densities, including single-family detached houses, low-rise patio villas, and high-rises of 20 or more stories. Also included, to serve a broader public, are a pedestrian-friendly town center with landmark office buildings, a variety of retail and restaurants, a conference center, a health club, a golf teaching center, and a hotel and resort. Since Luxe Hills' 667 acres must be developed in phases, it was crucial to establish a prestige identity and high market value from the outset. Contributing to this identity is the "Wan An" style adopted for the buildings, reflecting Asian character in forms and details, with a touch of Western relaxed elegance, using natural materials, warm colors, and patina finishes. Within this style, individual buildings are allowed a great deal of diversity. The layout of the site takes maximum advantages of appealing topographic features, such as, its many small hillocks, where groups of detached villas and other key building groups are located.

Dahlin Group

University Villages
Marina, California

While the closing of military installations can create immediate hardships to their localities, some can provide rare redevelopment opportunities. Such is the case for the site of Fort Ord near Monterey, across Highway 1 from a state beach. The 420-acre site consists of rolling hills with striking views of the Pacific, adjacent to the campus of California State University Monterey Bay. The proposed mixed-use community will include 1,237 residential units, 650,000 square feet of retail, and 650,000 square feet of office, research, and light industrial facilities. The plan is organized around two linear parks that cross at a central sculpture garden, one running east-west linking the university campus to an art district and the other forming a north-south link from the village center to a research/office area. Other large and small parks throughout the area provide community focal points. Traffic calming is achieved by narrow streets, narrowed intersections, and strategically located round-abouts. Readily accessible from Highway 1, the Village Center includes a mix of one-, two-, and three-story buildings, with a lively variety of storefronts, plus high-density and live/work housing. At the core of the Village Center, a one-acre Village Square forms a focal point for the village promenade, which evokes a classic main street atmosphere. The promenade normally includes parking along its wide, landscaped walks but can be cleared to accommodate special events. Homes will be of a "beach town character," largely clad with siding or shingles. Planting will be drought-tolerant native species, minimizing water demands. University Villages creates a true sustainable urban village.

Facing page, top: Site plan, with community area highlighted.

Facing page, bottom: Beachfront plaza.

Above: Live/work buildings.

Right: Houses of "beach town character" with native plantings.

Below right: Portion of research/office area.

Dahlin Group

Black Diamond
Pittsburg, California

A private development with support from the local redevelopment agency, Black Diamond is designed to bring residents back to downtown Pittsburg. The 7.3-acre mixed-use project will total 230,000 square feet, with 195 dwelling units occupying all of its upper floors. On portions of the street floors will be several sit-down restaurants, a coffee shop, a smoothie shop, retail shops, and boutiques — some widely known (Starbucks, Jamba Juice) and some unique — to create a downtown commercial destination. Street-level parking is wrapped with townhomes on three edges behind the mixed-use structures. Building design responds to the concerns of adjoining single-family neighborhoods and landmarks, including a historical church. The deeply sculpted exteriors are inspired by traditional styles found among historically significant structures in Pittsburg. The buildings recall numerous precedents, such as American Second Empire, with mansards and prominent window hoods, Art Deco, with smooth curves and low-relief details, and English Regency, with simplified Classical forms rendered in white stucco. A variety of canopies and arcades offer welcome weather protection along the first floor.

Below left and bottom: Architecturally varied mixed-use structures generating new downtown streetscapes.

Below: All-residential buildings with lower profiles, responding to nearby single-family areas.

Renderings: Digital Imaging Studio.

David M. Schwarz

1707 L Street NW
Suite 400
Washington, DC 20036
202.862.0777
202.331.0507 (Fax)
www.dmsas.com

201 Main Street
Suite 600
Fort Worth, TX 76102
817.339.1133

David M. Schwarz

David M. Schwarz

Fort Worth Master Plan
Fort Worth, Texas

Top: Computer model of Downtown.

Above and above right: Sundance East building with two floors of retail, one of offices.

Right: Sundance West apartment building with street-level retail.

Photography: Jim Hedrich/Hedrich Blessing.

Left: Sundance East restaurant building.

Below: Cinema entrance façade in mixed-use Sundance West block.

Right: Sundance East cinema building.

Since 1988, the firm has been involved in the revitalization of downtown Fort Worth, which had suffered economic setbacks after World War II, compounded by planning practices that only made the area less attractive and less safe. Beyond drawing up a sensitive and workable plan, the firm has been the architect for all of the key buildings and renovations discussed and illustrated on these pages, thus enjoying a rare opportunity to carry out its planning intentions and affect the character of a major urban core. The master plan initially focused on the 30 blocks around Sundance Square, but considered 150 blocks as significant context. Numerous parking lots, multistory garages, and set-back office buildings had interrupted commercial street life. Structures spanning streets had produced forbidding tunnels. The firm identified Third Street as one street with no major obstacles, recommending that it be restored to two-way traffic and lined with a mix of uses. To reintroduce residents essential to round-the-clock activity, the Sundance West apartments and the Sanger Lofts were sited on Third Street, initiating a process that has so far produced 1,200 units. Sundance West contains about 90 apartments, above an 11-screen AMC cinema, with street-level retail and underground parking. Sanger Lofts is a conversion of a 1928 department store, carried out under Department of Interior guidelines for historic preservation. The Sundance East project occupies a full city block, designed as a group of buildings in scale with historic neighbors. It comprises the Palace cinema, designed in the Moderne style plus a 28,000-square-foot Barnes & Noble store, some 26,000 square feet of dining places, and 11,000 square feet of office space. The expansion of the city's Central Library, including a new public entrance and lobby, extends the structure to help reconnect portions of the urban core. The Nancy Lee and Perry R. Bass Performing Hall for symphony, opera, and ballet companies — and the Van Cliburn International Piano Competition — seats up to 2,100 and is symmetrically laid out to fit its 200-foot-square city-block site, with its main entry at one corner to address the restaurants and shops of the Sundance

Top: Nancy Lee and Perry R. Bass Performing Hall.

Above: Bass Hall lobby.

Right: New main entrance to expanded Central Library.

Photography: Steve Hall/Hedrich Blessing, top and above, opposite top left and bottom tight; Jim Hedrich/Hedrich Blessing, right, opposite top right.

area. On a block next to Bass Hall, the Maddox-Muse Center houses recital, rehearsal, and other support spaces in a complex with the exterior appearance of four distinct structures. Two new office properties in the district, the Wells Fargo and Bank One Buildings, adopt the architectural character and the time-proven organization of historic commercial structures, built out to the sidewalk with retail at the street level. Today's Downtown Fort Worth has a wealth of commercial and residential resources that would hardly have seemed possible 15 years ago.

Below left: New Bank One Building.

Below: New Wells Fargo Building.

Bottom: Maddox-Muse Center, housing various performing arts functions.

David M. Schwarz

Southlake Town Square
Southlake, Texas

The master plan for a 135-acre site establishes a new downtown for a city that has never had one. The plan proposes a total build-out of 2.7 million square feet, of which the completed Phase 1 includes 300,000 square feet on 40 acres. For an area whose commercial development had been limited to strip commerce, the goal was a mixed-use, pedestrian-oriented environment that would appeal to residents countywide. Since most visitors would arrive by car, one objective was to accommodate cars without a negative effect on the pedestrian experience. The town has attracted numerous restaurants, shops, and offices, plus a town hall, and is the setting for many area celebrations and arts events. The mandated 21-meter right of way for its public streets was addressed with extra-wide sidewalks, generous plantings, angled parking, and necked-down intersections. Project is currently in Phase II which includes 150,000 square-foot retail/restaurant uses, a 14 screen cinema, a 200 room Hilton hotel, and structured parking for over 1600 spaces. Additionally, Phase I Residential is currently under construction and Phase II Residential is in the planning phase. The project has been praised by the Urban Land Institute and Landscape Architecture magazine, among others.

Top: Fountain in Town Square.
Far left: Sidewalk scene.
Left: Town Square, with bandstand/pavilion.
Below left: Street fronts seen from square.
Photography: Steve Hall/Hedrich Blessing.

David M. Schwarz

West Village
Dallas, Texas

West Village is designed as a hub for Dallas's Uptown District, which has evolved over the past decade as a premier residential, shopping, and entertainment area. The development's 6.75-acre site is favored with exceptional public transportation. It is two blocks west of the Dallas Area Rapid Transit's (DART's) CityPlace station and is ringed by the Uptown Dallas Trolley's terminal loop. The project includes 178 luxury apartments and 125,000 square feet of high-end retail. Building heights range from one to four stories, with retail on the lower levels, residences above. Parking is provided by spaces along its streets and in an 800-space garage, hidden by mixed-use structures. Building heights and architectural treatments vary in response to neighboring development, with a loftlike structure facing the site of high-rise structures planned near CityPlace station, lower buildings facing lower-density areas, and a Mediterranean character along Cole Avenue, where this style was already prevalent.

Top: Retail street scene.

Above: View of village along one perimeter.

Above right: Main entrance to Cole Avenue apartments.

Right: Mixed-use frontage reflecting Mediterranean style of older neighbors.

Photography: Steve Hall/Hedrich Blessing.

David M. Schwarz

Parker Square
Flower Mound, Texas

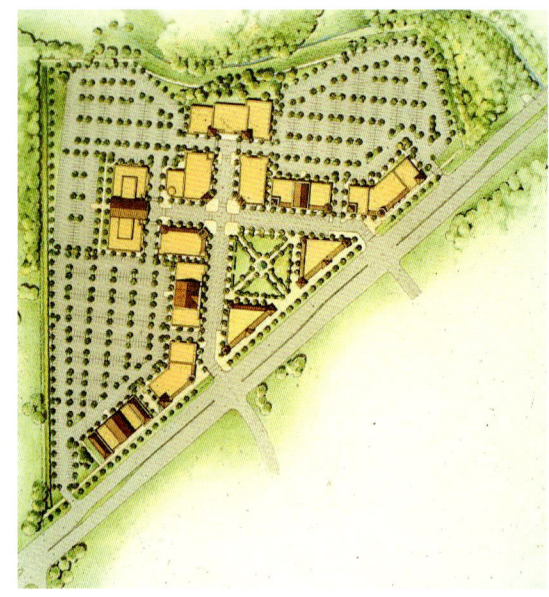

Before Parker Square was built, Flower Mound was a suburban locale — near Dallas-Fort Worth Airport — without any sense of neighborhood or community focus. For a 24-acre site previously approved for an office park with pad-site retail/restaurant along the highway, the architects worked with a new owner and town officials to produce zoning and development guidelines for a more pedestrian-friendly, mixed-use development that can serve as a focus for Flower Mound as a whole. Because site limitations and area demographics reduced the likelihood of large-scale tenants, the project is geared to neighborhood commercial tenants and smaller office users. Offices are located on the second level around the central square. Buildings are carefully laid out to screen the majority of the site's parking areas from the approach roads, the square, and the pedestrian streets. Building facades feature brick details in creative interpretations of regional Main Street architecture.

Above: View from square.
Above left: Plan of development.
Left: Square with pavilion.
Below left: Façade detail of domed building.
Below: Typical offices-over-retail buildings.
Photography: Steve Hall/Hedrich Blessing.

Dougherty Schroeder & Associates, Inc.

211 Perimeter Center Parkway
Suite 900
Atlanta, GA 30346
770.650.7774
770.650.7708 (Fax)
kdougherty@dsaarch.net
www.dsaarch.net

Dougherty Schroeder & Associates, Inc.

Dougherty Schroeder & Associates, Inc.

Destin Commons
Destin, Florida

The "New Urbanism" exemplified in the Florida Panhandle communities of Seaside and Rosemary Beach can now be experienced in the larger city of Destin, nearby. This retail and office development brings the same kind of pedestrian-friendly atmosphere and civic symbolism to a new commercial hub for the Emerald Coast. It is designed to appeal to a wide spectrum of the public. The 56-acre project includes 420,000 square feet of retail space, with 70,000 square feet of office at second-floor level. The generous use of natural stone, brick pavers, roof tiles, slate, and other traditional, durable materials lends the project a substantial feeling not common to commercial developments in the area. Landscaping is relatively formal, and outdoor lighting is of the subtly modulated kind found in high-end resorts. The project is the first phase of a larger development to include additional retail and office as well as an upper level hotel.

Top: Fountain plaza.

Left: Evening view of shops.

Above: Focal fountain.

Below: Destin Commons elevations.

Facing page: Two signature towers.

Photography: Craig Tanner and Dennis O'Kane.

Dougherty Schroeder & Associates, Inc.

Pinnacle Hills Promenade
Rogers, Arkansas

Left: Plan of development.
Left center: two views of Central Park.
Below: Street scene.
Bottom: Typical building elevations.

The goal here was to create a new downtown for Rogers, which is experiencing explosive growth as the headquarters of Wal-Mart, the world's largest corporation. The development will include 1.1 million square feet of retail, a theater, and second-level office spaces, organized around a 1.33-acre park of urban character. The architects' first inclination was to base the design of buildings on the appealing 19th-Century precedents represented in nearby communities such as Eureka Springs and Fayetteville. Discussions with the developers, however, soon convinced them that the city's extensive international connections — attracting business people from all over the world — made a less regional, predominantly Modern design vocabulary more appropriate.

Dougherty Schroeder & Associates, Inc.

The Avenue East Cobb
Atlanta, Georgia

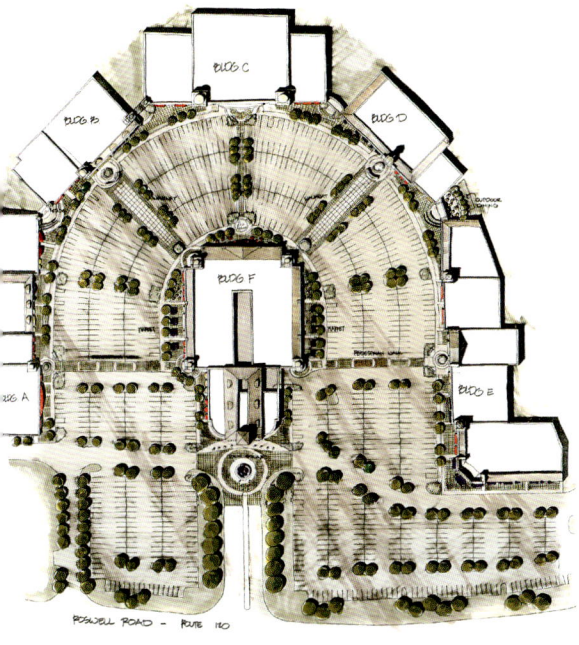

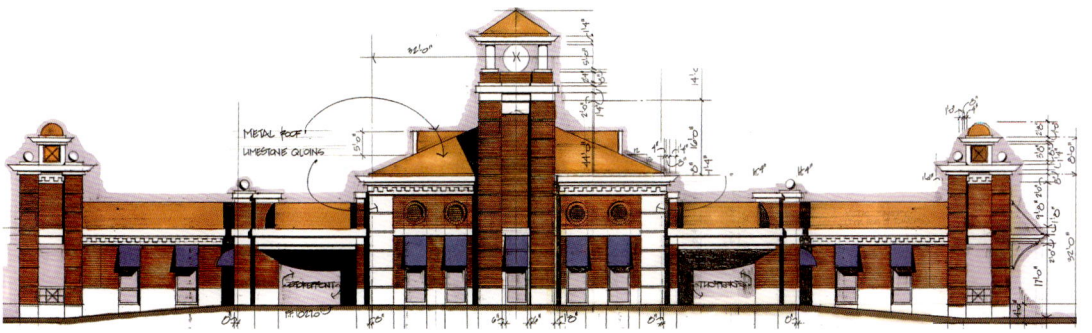

Far left: Horseshoe plan of center.
Left: Three views of shopping environment.
Bottom: Characteristic elevation, showing central tower and wings.
Photography: Dennis O'Kane,
A.O.R.: CMH, Inc.

This 240,000-square-foot lifestyle center in Atlanta's Cobb County suburbs established a signature look for subsequent Avenue retail projects. For this site, local zoning would not permit inclusion of residential functions. The layout and architectural design recall the pattern of an Antebellum town square, with a centrally located "transportation building" resembling a 19th-Century railroad depot, surrounded by lower structures. Lushly landscaped "pedestrian boulevards" linking the buildings modulate the passage from car to stores and encourage strolling around the project. Storefronts for some 50 retailers, many nationally known, are urged to display individuality within the unifying framework composed of red brick and traditional molding and belt courses. Fine materials such as tumbled brick pavers and standing-seam metal roofs maintain the center's appeal for both retail tenants and their customers.

Dougherty Schroeder & Associates, Inc.

The Forum at Sunnyvale
Sunnyvale, California

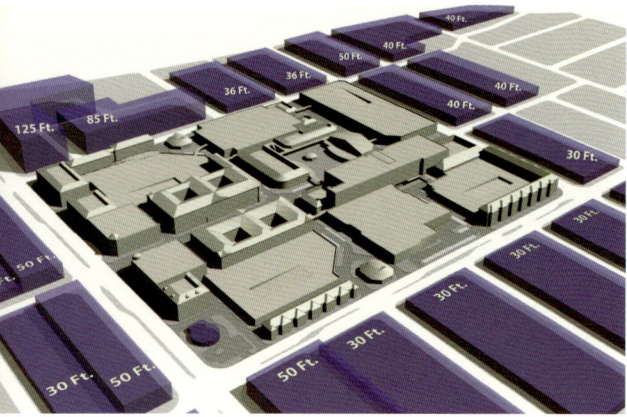

Left: Massing model.

Below left: View from the main thoroughfare.

Bottom: Representative building elevations.

Facing page top: Typical street view.

Facing page bottom: Public open space.

Here the architects were challenged to recreate an urban core that had been decimated in the 1970's by the displacement of six city blocks by an in-town shopping mall. Their design replaces the mall with a more traditional layout of streets lined with mixed-use structures, including first-floor shops, a department store, and a three-level theater, with offices and housing units on upper floors. In accordance with local planning and zoning decisions, there will be 1.1 million square feet of retail, 250,000 square feet of offices, 330 housing units, and 5,000 structured parking spaces, along with 1.5 acres of public open space. Building surfaces will include stone, brick, and stucco, with slate and tile roofs. Generous landscaping will include some existing 100-foot redwoods and Monterey cypresses.

135

Dougherty Schroeder & Associates, Inc.

Gulf Coast Town Center
Fort Myers, Florida

Spanish Colonial building fronts will surround formal plazas and subtropical water gardens in this 1,700,000-square-foot retail development in suburban Fort Myers. Its axial plan will focus on a column-topped circular fountain and a variety of garden towers, pergolas, and pavilions, with vistas delineated by tall palms. Generous arcades and awnings will protect patrons from sun and weather. Stuccoed retail structures will be topped with a variety of tile roofs. Signage will be restrained. Occupying approximately 204 acres, the complex will include a 770,000-square-foot life style center "Main Street" development, with restaurants and three anchor department stores, as well as a perimeter Big Box retailer and a cineplex.

Above: Fountain and main axis.
Above left: Aerial rendering.
Left: Central water garden.
Bottom: Building elevations.
Model: Pacificom Multimedia, Inc.

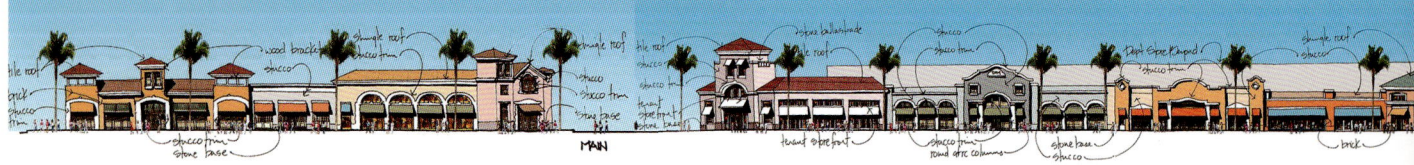

Duany Plater-Zyberk & Company

1023 SW 25th Avenue
Miami, FL 33135
305.644.1023
305.644.1021 (Fax)
www.dpz.com

Duany Plater-Zyberk & Company

Duany Plater-Zyberk & Company

Tannin
Orange Beach, Alabama

The plan for this coastal village on the Gulf of Mexico responds to several challenging characteristics of the site, including its irregular boundaries and the linear dune and wetland formation that run oblique to the main highway. Wetlands within the site have been transformed into linear lakes, while the swamp at its edges has been left wild, framed by residential lots and a public pavilion terminating the main boulevard. A town square at the highway gathers the public buildings including a village hall, post office, and regional fire station. Tannin's urban regulations prescribe the physical characteristics of traditional Southern building types, and architectural regulations specify construction materials and techniques that are economical and found in the local vernacular.

Top: Typical Tannin cottage with front porch, front yard of native vegetation, and picket fence along street.

Above: Houses along a lake.

Above left: Street leading to swimming pool pavilion.

Far left: Site plan.

Left: Public fountain with distinctive pump house.

Photography: George Gounares.

Duany Plater-Zyberk & Company   Amelia Park
Fernandina Beach, Florida

An exemplary infill neighborhood, the 106-acre Amelia Park is at the heart of a historic coastal city. Located close to existing community facilities, the development offers convenient housing for a wide range of ages and lifestyles, including single-family houses, rowhouses and apartments over shops. The master plan draws upon some of the best-loved towns in the region, particularly the city's own historic downtown. Understanding the value of this example, the city adopted the proposed Amelia Park Plans and Codes as part of a planned unit development ordinance and established a Community Development District to facilitate self-governance and financing. Now 55 percent completed, the neighborhood will contain about 450 residential units, 70,000 square feet of commercial space, 25,000 square feet of professional offices, and various public amenities, including a YMCA on a lakeside site.

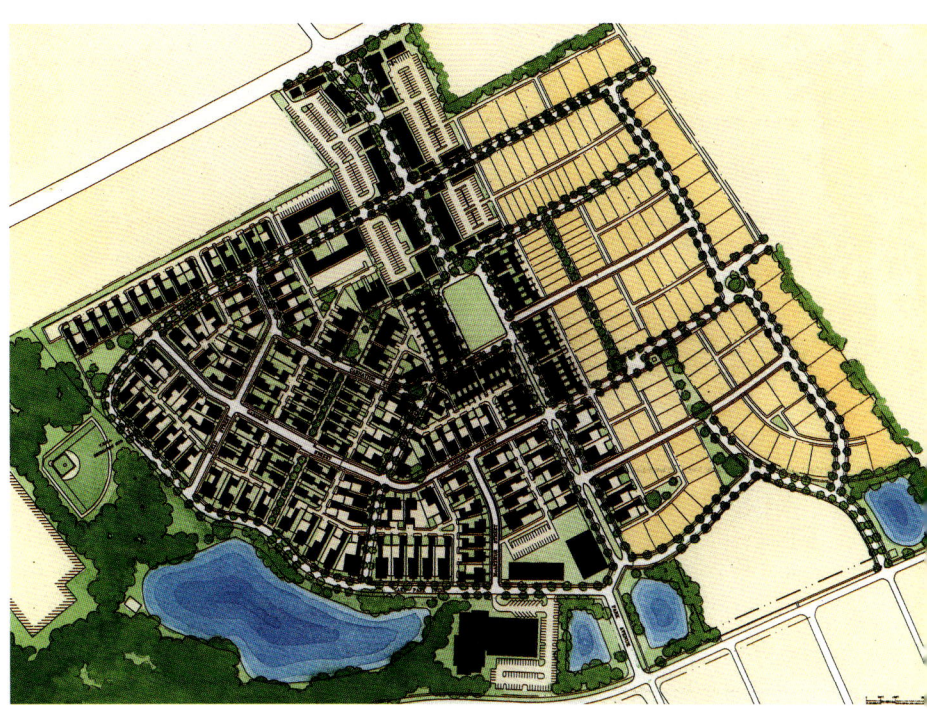

Top: Neighborhood plan, with streets and parks linked to adjacent areas.

Above and above left: Rowhouses along typical streets.

Left and far left: Variety of residential types from large homes to small cottages.

Photography: DPZ, Oscar Machado.

Duany Plater-Zyberk & Company

I'On
Mount Pleasant, South Carolina

Left: Charrette Master Plan.
Below left: A series of townhouses.
Below right: Porch houses fronting one of the man-made lakes.
Bottom left: Commercial buildings at central square.
Bottom right: Houses overlooking East Lake across public street, diverted to accommodate existing trees.
Photography: DPZ.

The 243-acre site near Charleston is bordered by a marshy creek to the north and a rural thoroughfare to the south, with residential subdivisions to either side. The property features three manmade lakes, a historic graveyard, and a monument to local hero Jacob Bond I'On. The master plan locates a village center close to the main road, with commercial uses around a central square. Neighborhood centers include one on the creek, which includes a community boat landing and a meeting hall. The smallest of the lakes is left as a wildlife sanctuary, and the others are surrounded by porch houses. Rowhouses and single houses are located in the more central areas, with porch houses in the more rural and marsh-front sites. The master plan was designed in collaboration with Dover-Kohl & Partners.

Duany Plater-Zyberk & Company

Habersham
Beaufort, South Carolina

Above: Town plan.
Right: Edge along wetlands.
Below right: Single-family houses among mature trees.
Bottom left: Rowhouses.
Bottom right: Boardwalk over tidal marshes.
Photography: DPZ.

Located on the water just eight minutes from historic downtown Beaufort, the new town of Habersham provides the Low Country with an alternative to suburban sprawl. The town's architecture respects the expert methods employed in traditional designs for ventilation and cooling and mandates them in the codes that accompany the plan. Porches within conversation distance of the sidewalk encourage socializing. An assortment of residential building types encourages a diverse population and permits a single development to address a variety of market segments. A neighborhood center at the main entrance to Habersham is designed to grow as demand increases, with buildings of the historic Main Street type, with office or residential spaces above street-level retail. The plan was a collaborative effort of DPZ with Steven Fuller Design Traditions, Moser Design Group, and Historical Concepts.

Duany Plater-Zyberk & Company

Rosemary Beach
Panama City, Florida

Located on the northwest Florida coast seven miles east of the famous 1980s New Urbanist community of Seaside, Rosemary Beach represents 15 years of growth in Duany Plater-Zyberk's urban design experience. The community straddles the coastal highway, which crosses the town through a linear park. The design team distinguished Rosemary Beach from Seaside in several ways. They inserted mid-block vehicular alleys, permitting pedestrian boardwalk promenades leading to the beach, and they created civic spaces at the water's edge to emphasize the connection to the Gulf. About half of the alley-fronting garages are topped by granny flats. The community's center features live-work and arcade buildings, a town hall, and a post office. In contrast to Seaside's Key West vernacular, the architecture here is based on models from the Caribbean and St. Augustine, with stuccoed or shingled walls, cantilevered balconies, and shuttered sleeping porches. Now 70 percent completed, the community includes 470 residential units ranging from studio flats to six-bedroom cottages and a substantial portion of the allotted 97,000 square feet of commercial space.

Above: Master Plan, with coastal highway in linear green swath.

Below: Beach-bound boardwalk between house fronts.

Right, top to bottom: Variety of houses along boardwalk promenade; community pool and beach house; Caribbean-inspired house details; town hall and post office with walled courtyard.

Right: Mixed-use buildings in a variety of types.

Below: Town center, Barrett Square, with town hall at one end.

Photography: DPZ.

Duany Plater-Zyberk & Company

Alys Beach
Panama City, Florida

Top left: Master plan, with Gulf to south, highway green belt through center, and wetlands to north.

Top right: Complex with walled courtyards, encouraged in plan.

Above left: Exterior view of the sales office featuring a distinctive Bermuda-style chimney and scalloped gable parapets.

Above right: Front entry of the sales office with a pedestrian friendly planter that doubles as a bench.

Left: Courtyard and loggia of sales office.

Bottom left: Detail view of Bermuda-inspired rooftops.

Photography: Alys Beach.

A 160-acre resort community in the tradition of nearby Seaside and Rosemary Beach, Alys Beach is designed as a pedestrian-friendly community in harmony with nature and inspired by the graceful vocabulary of Bermuda's architecture. The planned 500 residential units range from above-the-shop apartments to family compounds, and the community will include 170,000 square feet of commercial and office space. Residents' front doors face footpaths tracing a network through the development. Courts on the interior of many blocks hide automobiles from view. Environmentally features include the use of permeable paving, natural storm drainage, wind energy, and cisterns. The community's white masonry walls and roofs, based on Bermuda models, contribute to energy efficiency.

Elkus Manfredi Architects

300 A Street
Boston, MA 02210
617.426.1300
617.426.7502 (Fax)
www.elkus-manfredi.com
info@elkus-manfredi.com

Elkus Manfredi Architects

Elkus Manfredi Architects

35 and 40 Landsdowne Street
University Park at MIT
Cambridge, Massachusetts

Far left: Site plan.

Below left: Stone monoliths included in artwork by David Phillips.

Below: Quadrangle defined by 35 and 40 Landsdowne, left and center in photo.

Right: Entrance to 40 Landsdowne.

Far right: 35 Landsdowne and Quadrangle.

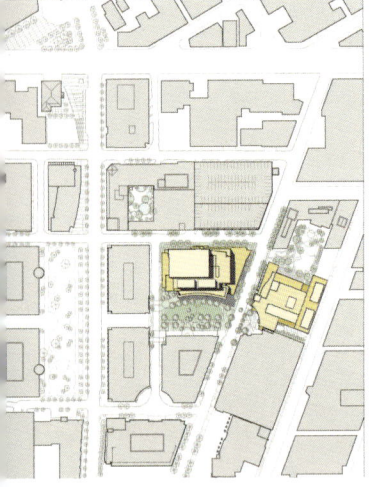

Historically, the intense chemistry lab has been low-rise, internally oriented and suburban. Driven by the desire for proximity to the resources of the Massachusetts Institute of Technology, 35 and 40 Landsdowne Street, on the "campus" of University Park at MIT, represent a new laboratory prototype: urban, high-rise, high-hazard-rated facilities for intense chemistry and biology use. Together, the buildings fulfill important urban roles: shaping open spaces, defining the public realm of surrounding streets, and energizing the adjacent sidewalks and park. Combined as the headquarters of Millennium Pharmaceuticals, one of the country's leading biopharmaceutical companies, the buildings serve as research facilities and corporate offices. The buildings

include a lecture hall and a cafeteria, as well as labs and offices. Design goals for the buildings were as follows: provide maximum flexibility for a rapidly changing science; encourage planned and unplanned interaction between scientists of different disciplines; draw employees from other buildings to foster interactions across the company. A commitment to sustainable design was implemented through a number of design strategies: reuse of a brownfield site; building-wide air-side management; lab waste management; maximum daylighting benefits; high-efficiency lighting and light controls; water conservation and gray water recycling; design for long-term adaptive reuse with minimal demolition and discarding of building materials; integrated systems commissioning. The 35 Landsdowne Street building has a total floor area of 220,000 square feet, and 40 Landsdowne, 225,000 square feet. The half-acre Landsdowne Quadrangle, defined by the buildings, features a 40-foot by 100-foot artwork by David Philips, which consists of granite monoliths up to 9 feet high, along with bronze inserts paving, granite and lighting installations.

Facing page: Entrance canopy, 35 Landsdowne.

Left: Cafeteria at 35.

Below left: Lounges at 40 Landsdowne.

Below: Overall view of 40.

Bottom: Night view of 40.

Photography: Justin Maconochie (35 Landsdowne), Bruce T. Martin (40).

Elkus Manfredi Architects

100 Cambridge Street/Bowdoin Place
Boston, Massachusetts

Far left: Site plan.

Near left: Entrance to Saltonstall Plaza and Garden of Peace, with Ibis Ascending sculpture by Judy Kensley Mckie.

Below left: Wall sculpture Freedom One by Howard F. Elkus over office building escalator.

Below: 100 Cambridge Street building lobby.

Facing page: Saltonstall Plaza, with Leverett Saltonstall building at right.

Rehabilitation of the 22-story Leverett Saltonstall State Office Building has yielded a dividend by mending the fabric of a historic neighborhood once divided by the tower's vacant plaza. When the Massachusetts legislature invited development proposals for the defunct structure, Elkus Manfredi, members of a team assembled by MassDevelopment, produced a winning scheme that wrapped five-story mixed-use buildings around three sides of the once-forbidding plaza. The proposal achieved much more than the state anticipated by including private components that carried much of the project's cost. Residential buildings and street-level retail and below-grade parking now line Bowdoin and Cambridge Streets, which are essential components of the Beacon Hill cityscape. The renovated office building now presents two distinct images and two addresses, successfully combining very different public and private identities and functions. The needs for building access and public space have been met by transforming the remaining plaza area into two parks: the Garden of

Peace, dedicated to victims of violence, and Saltonstall Plaza, honoring a public figure's legacy of service. Permanent art installations engage users and enhance the project's public areas. The completed project includes 133,000 square feet of residential construction and 34,500 square feet of retail, along with the 565,000-square-foot office building.

Above: Residential buildings along Bowdoin Street, with renovated Saltonstall building behind them.

Below: 100 Cambridge Street entrance to office building, with 60-foot Sol Lewitt mural, flanked by mixed residential/retail structures.

Below right: New mixed-use buildings along Cambridge Street, with Saltonstall building in background.

Photography: Woodruff Brown Photography.

ELS
Architecture and Urban Design

2040 Addison Street
Berkeley, CA 94704
510.549.2929
510.843.3304 (Fax)
info@elsarch.com
www.elsarch.com

ELS Architecture and Urban Design

California Theatre
San Jose, California

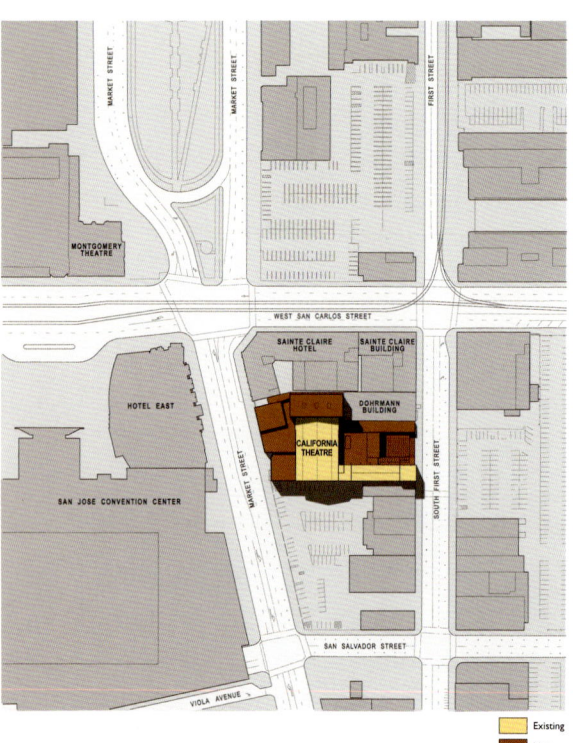

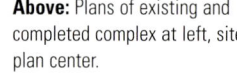
Existing
New

Above: Plans of existing and completed complex at left, site plan center.

Left: Addition with new entrance and support spaces opposite convention center.

Below left: New courtyard.

Below right Historic façade.

Above right: Renovated theater interior.

Facing page: Part of restored lobby.

Photography: Marco Zecchin (exterior and lobby), Rudolph & Sletten (courtyard), Tim Griffith (theater interior and historic façade).

With a combination of city and nonprofit funding, a historic movie theater has been renovated and expanded as a performing arts facility. The original 1927 building, presenting an exuberant display of ornamentation, has been carefully restored. Alterations to the theater interior include expansion of the orchestra pit and installation of a pit elevator, new lighting and sound systems, and re-raking of rear orchestra and balcony seating to improve sight lines. For required seismic upgrading, new supports have been added on the exterior of the structure to avoid disturbing historic interiors. To accommodate the needs of touring companies, opera groups in particular, the stage house has been expanded, and that required relocation of a service alley serving adjacent buildings on the block. Other additions provide for dressing rooms, a green room, a conference room, and a large rehearsal space, restrooms, truck loading, and other back-of-house support spaces. Selected ancillary facilities added to the existing theatre formed a new outdoor courtyard along the main street, accessible to patrons at intermission and to the general public at non-performance times. A second main entrance and lobby have been added across the street from the city's convention center to facilitate the theater's use in conjunction with meetings.

ELS Architecture and Urban Design

Church Street Plaza
Evanston, Illinois

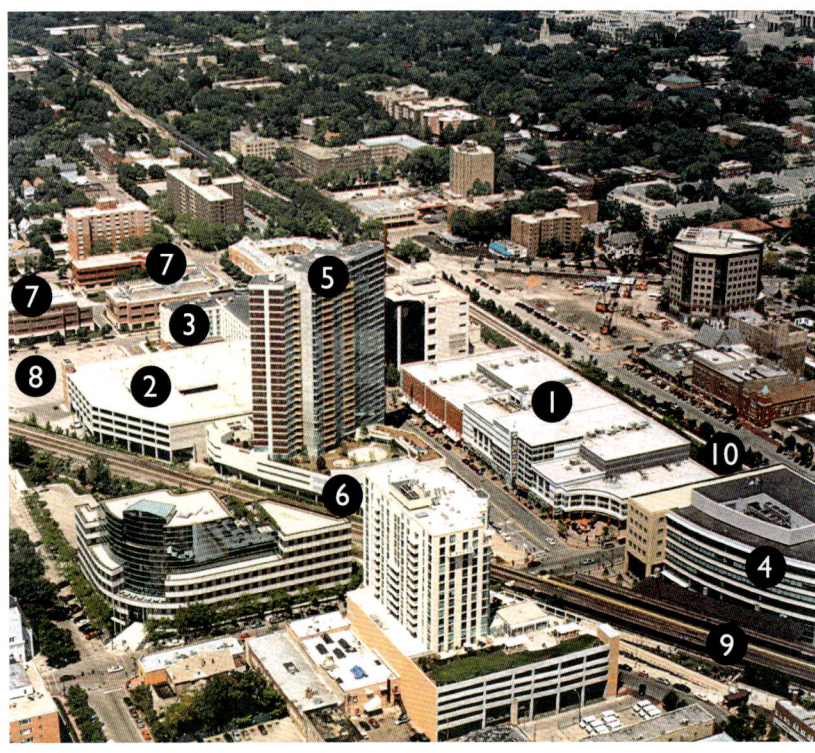

Left: Aerial view of master plan development
1. Main Pavilion retail/entertainment
2. Public garage
3. Hotel
4. Office/retail building
5. Residential condominium
6. Retail
7. Northwestern University
8. Existing Saturday farmers' market
9. Metra Rail
10. CTA Rail

Below: Redeveloped Maple Avenue, with Main Pavilion on right, retail base and residential condominium on left.

Facing page: Two views of main pavilion, with street-floor retail and second-floor cinema. ELS--design architect, DeStefano + Partners--executive architect.

Photography: ELS (Maple Avenue), Timothy Hursley (Main Pavilion).

On a 7.5-acre tract between Northwestern University and Evanston's historic center, this development initiated the city's revival as an entertainment and residential center. Downtown eating and drinking places have increased from almost none to 85, new housing and a hotel have generated round-the-clock activity. ELS developed the master plan for the entire site and were the design architects for the main pavilion and parking garage. Strategically located between the garage and two regional transit stops, the pavilion houses retail and restaurants at street level with an 18-screen multiplex on its second level, its lobbies in dramatic glass-walled volumes. Retail and entertainment facilities for the whole development total 174,000 square feet. Other components include a 178-unit hotel; 190,000 square feet of office space; 207 housing units; and a 1,400-car garage. With original ownership of the site split between the university and the city, the redevelopment process was complex. After a lengthy selection procedure, the city agreed on one master developer, teamed with ELS, who offered a pedestrian-oriented plan, rather than an introverted mall, and proposed the second-level cinemas above retail.

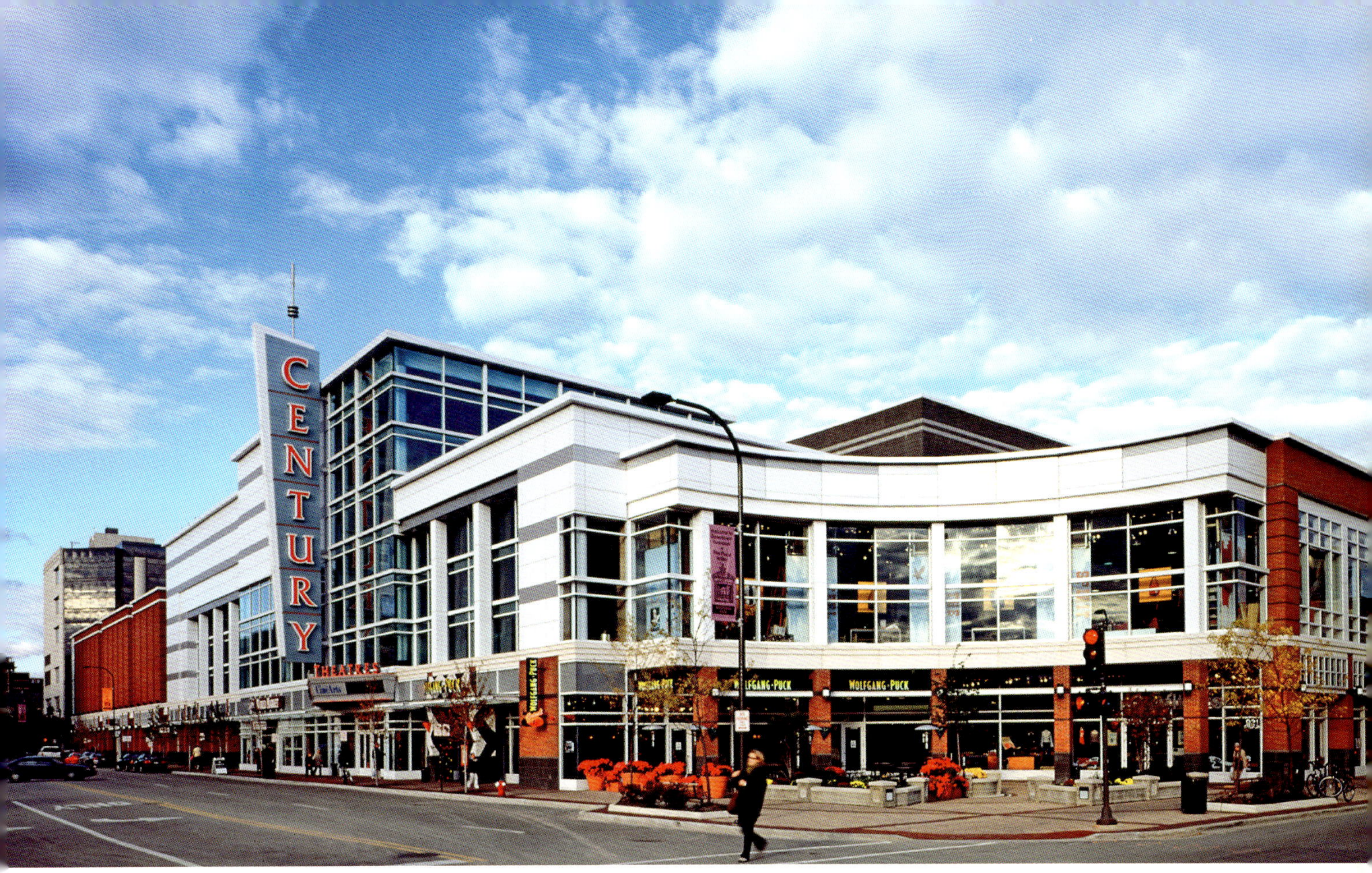

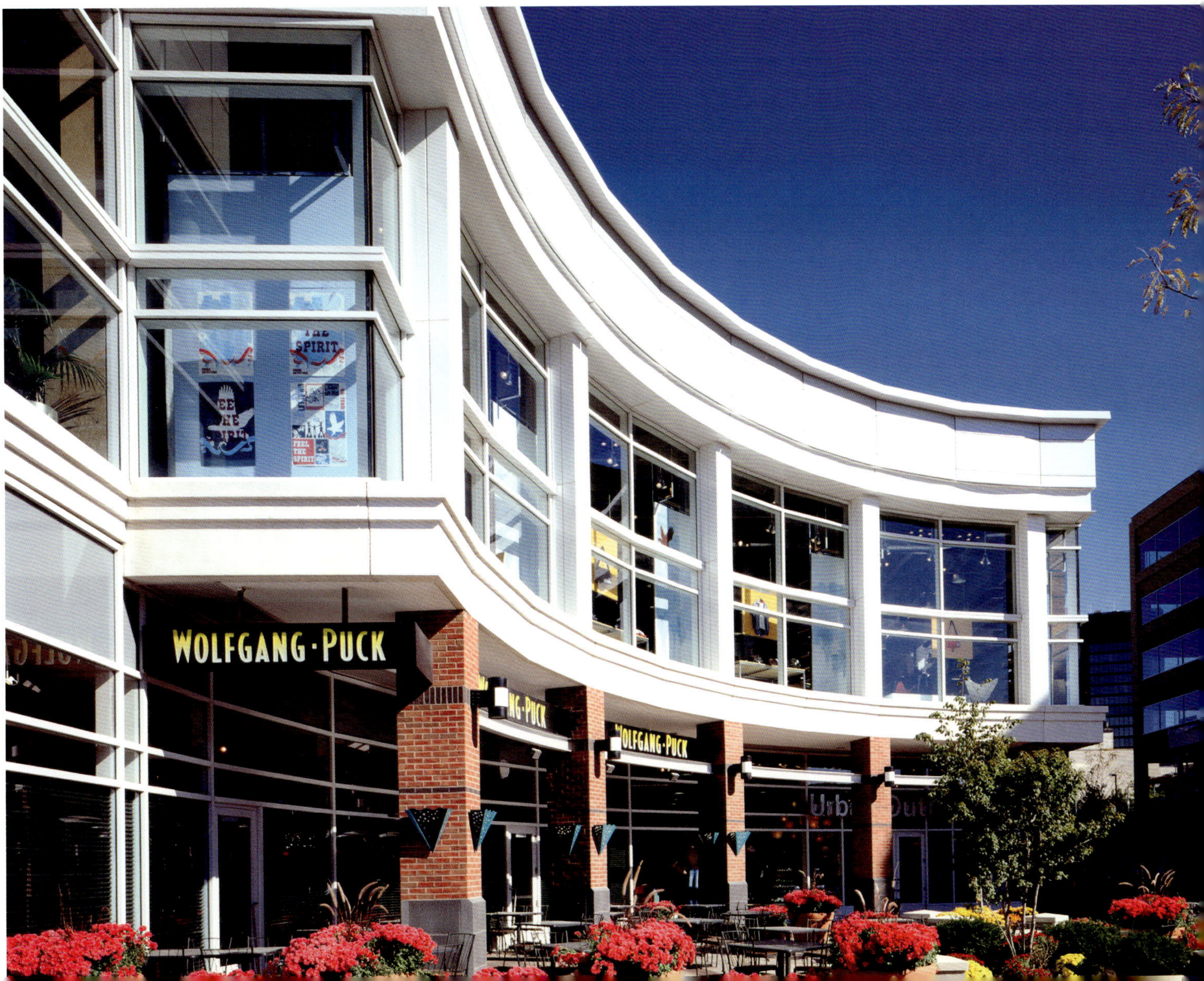

ELS Architecture and Urban Design

The Village of Merrick Park
Coral Gables, Florida

Left: Central garden, with main retail arcade.

Below: Master plan.

1 Miraflores Arcade 3-level retail building

2 2-level retail pavilions

3 Department stores

4 2,000-car parking structure

5 Central garden

6 Residential buildings

7 Office building/parking structure

8 Rail line

Facing page, top: Two intimate fountain courts.

Facing page, middle right: Landscaped stairs connect housing to garden.

Facing page, bottom right: Central garden, with paired residential buildings at far end.

Photography: ELS (plan and facing page, lower photo), SWA (upper top photos).

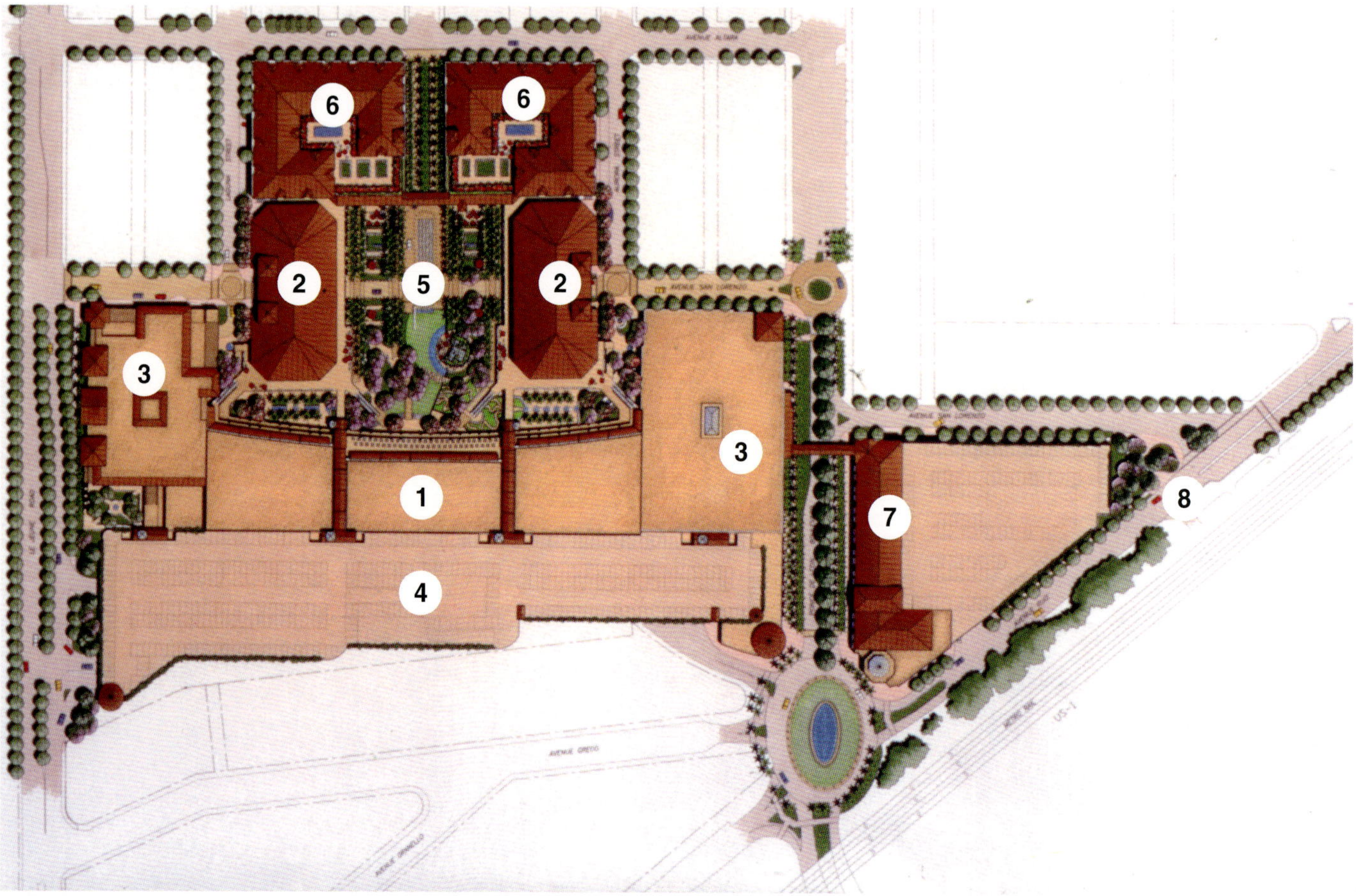

This mixed-use development organizes retail, entertainment, and residential uses around a central, pedestrian-scaled garden. Working closely with the developer, the architects designed retail structures that define a palm-lined green as a setting for round-the-clock, urbane activities. The main three-story retail building, the Miraflores Arcade, fronts on the central garden, which is flanked by a pair of two-story retail pavilions. At the end of the garden facing the main arcade are paired residential buildings. The Miraflores Arcade contains 280,000 square feet of retail, with a three-level Nordstrom store at one end and a three-level Neiman Marcus at the other. The two retail pavilions total 130,000 square feet. The Mediterranean-inspired architecture builds on Coral Gables' design guidelines that maintain a style established in the 1920s. The development opens itself up to the existing community with public streets running through it, following the historic city plan. The subtropical climate allows all circulation to be open-air, with a series of balconies, terraces, arcades, and loggias. Fountains and shade trees provide an appealing environment for outdoor eating and relaxation. A 2,000-car parking structure adjoins the main retail arcade. Perkins & Will designed the two residential buildings with 120 units and, adjacent to the roundabout, a 140,000 sf office building with parking structure.

ELS Architecture and Urban Design

The City of Sunnyvale Downtown Design Plan
Sunnyvale, California

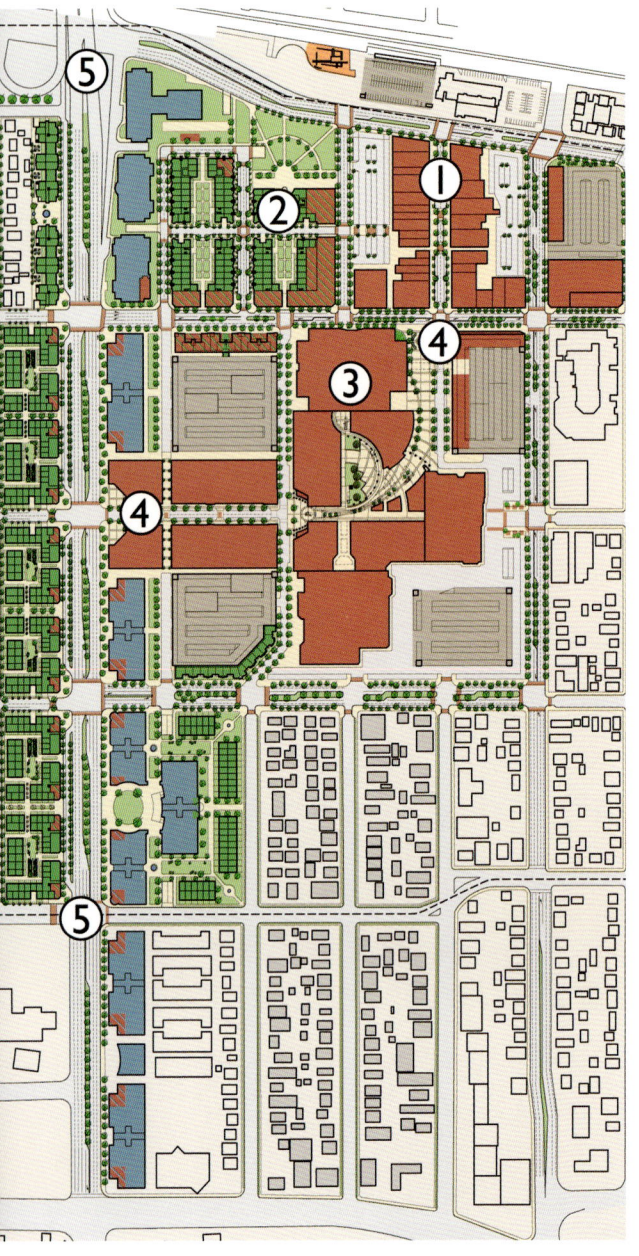

Left: Recommendations of Downtown Design Plan.

1 Defend and enhance historic districts

2 Add high-density housing downtown

3 Transform mall into open-air, mixed-use district

4 Reconnect historic street grid

5 Create a central boulevard lined with new offices and housing

(Red = retail; blue = offices; green = housing; gray = garages; cross-hatching for mixed uses.)

Below: Portion of new boulevard, lined with mid-rise office and residential buildings, lower housing bordering existing neighborhoods.

Bottom: Housing with some street-level retail around new park adjacent to train station.

Photography: Gerald Ratto (model).

Nine blocks at the center of Sunnyvale were demolished in 1977 for a shopping mall and surface parking, leaving only one block of the traditional shopping street intact. As new development pressures rose, the city commissioned ELS, working with an economist and a Downtown Stakeholders Advisory Committee, for an updated design plan covering 18 blocks. Workshops and public meetings helped shape the plan. Key goals are to reinforce the surviving commercial block and surrounding neighborhoods and to reestablish links in the disrupted city street grid. Anticipating renovation of the retail mall, the plan proposes transforming it into an open-air, mixed-use district. An existing arterial would be transformed into a central boulevard, lined with mid-rise housing and offices, a gateway to the downtown.

Field Paoli

150 California Street
7th Floor
San Francisco, CA 94111
415.788.6606
415.788.6650 (Fax)
architects@fieldpaoli.com
www.fieldpaoli.com

Field Paoli Beverly Canon Mixed-Use Retail
Beverly Hills, California

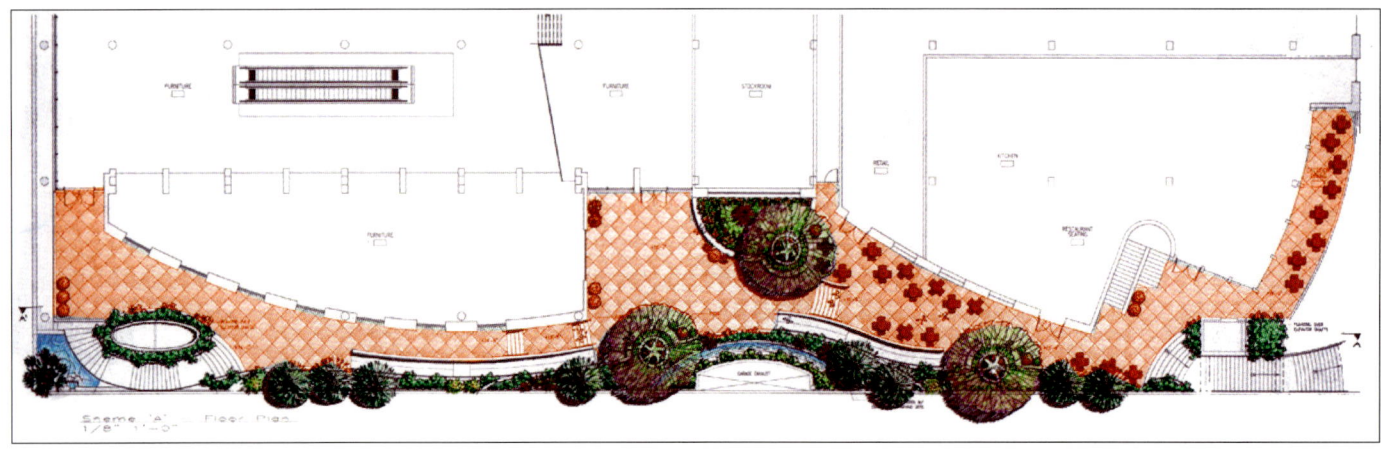

Located in one of the world's most fashionable shopping districts, this innovative mixed-use project was developed by the City of Beverly Hills to support the City's Strategic Plan. With a new anchor store and over 400 parking spaces, the project reinforces existing retail uses at the northern edge of the "Golden Triangle." Situated on a 48,000-square-foot infill property fronting Beverly Drive on one side and Canon Drive on the other, the complex includes 70,000 square feet of retail, 20,000 square feet of office space, and a four-level underground parking structure. Fulfilling the City's plan for enhanced pedestrian circulation, a landscaped plaza provides a continuous raised walkway over the mid-block service alley. Accessed by a number of stairways, ramps and elevators, this intimate urban oasis is open to the public at all times. The clean-cut geometric forms of the building are composed of limestone, plaster, glass and metal, creating a contemporary expression that reflects the optimism and self-confidence of this forward-looking city.

Facing page, top: Elevated walkway linking Beverly Drive and Canon Drive.

Facing page, bottom: Plan of walkway.

Above: Canon Drive elevation.

Above left: Fountain detail in walkway.

Below left: Diagram showing organization of project, with walkway crossing central alley.

Below: Beverly Drive elevation, showing elevator cylinder.

Photography: Jay Graham.

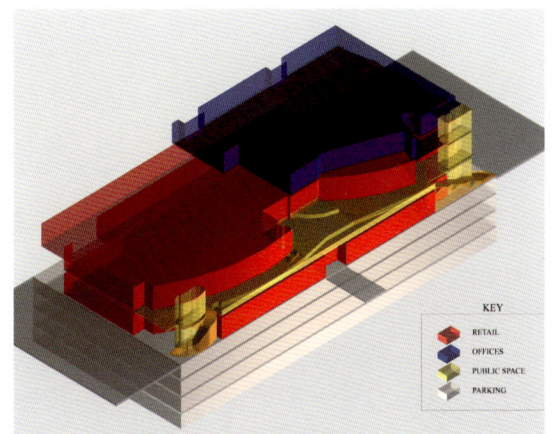

Field Paoli

On Broadway
Downtown Redwood City, California

Right: Cineplex entrance with flanking storefronts.

Below: View of proposed outdoor dining.

Below right: Cylindrical corner entrance at retail anchor.

Building a 160,000-square-foot retail/entertainment destination in the historic core of Redwood City required design sensitivity, community support, and resourceful teamwork. Landmarks such as the 1921 fire station (now a library), the 1930 courthouse, the 1912 Sequoia Hotel, and the 1928 Fox Theater contribute to a unique downtown environment. The Post Office and the City Hall are directly across the street from this full-city-block development.

To fit into this context, the cinemas are located on the second floor, allowing for the introduction of street level retail on the ground floor. Wrapping the theater with a glazed exit corridor and giving the ground floor retail a two-story expression has eliminated the blank walls typical of many of today's multiplexes. To produce a scale compatible with surrounding buildings, building facades are broken into discrete elements, with strong recesses and projections. Exterior details recall the Art Deco and Art Moderne commercial buildings of the 1930s, which were particularly congenial to storefronts and discreet signage. By bringing in new types of businesses, as well as the cinema complex, the project aims to increase the number and duration of visits to Downtown. With its close proximity to the Redwood City CalTrain station, On Broadway is a prime example of transit-oriented urban development.

Above right: Site plan, showing neighboring landmarks, with new development in red.

Right: "Flatiron" corner with canopied entrance.

Field Paoli

Victoria Gardens
Rancho Cucamonga, California

1850s

1900s

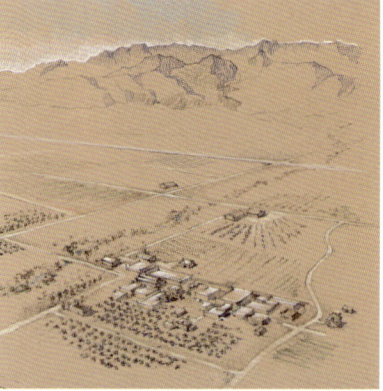
1950s

Today

Top: Conceptual sketches of community growth.

Right: Aerial view of Victoria Gardens.

Below left: New downtown.

Below right: Evening view of retail street.

Working closely with Forest City Development and the City of Rancho Cucamonga, Field Paoli created a Master Plan for a new 120-acre downtown. The open air town center is laid out with a new street grid designed to accommodate retail, housing and office uses, as well as a library and cultural arts center. During the master planning process, several architectural firms joined Field Paoli to provide distinct building designs for the initial retail phase of the project. The result is a rich aesthetic variety characteristic of evolving downtown streetscapes. In the central core, street widths were narrowed to slow traffic and provide an intimate shopping and dining environment. Specific sub-districts were identified for high-end fashion, lifestyle tenants and cafés around a central square. Smaller pocket parks and plazas enliven the street edges and create additional public gathering spaces. At the periphery, the Master Plan identifies surface parking areas that can be transformed into pedestrian-oriented city blocks as this Southern California city continues to grow.

Above: Public area off street.
Above left: Exterior window and façade detail.
Left: Pedestrian activity on street.
Below: Pocket plaza at night.
Photography: Jay Graham.

Field Paoli

The Streets of Tanasbourne
Hillsboro, Oregon

Above: Main Street.

Left: Central plaza at bend on Main Street.

Below left: Terraced gardens rising from central plaza.

Below: Gateway at perimeter boulevard.

Bottom: Site plan.

Located 10 miles west of Portland, The Streets of Tanasbourne was planned as a new retail neighborhood with a department store anchor and over 40 stores and restaurants totaling approximately 385,000 square feet. The project represents a more urban approach to suburban retail, with multi-level stores wrapping mid-block structured parking and fronting onto a curving "Main Street." Continuing Portland's long-standing tradition of pedestrian open space, a network of landscaped outdoor plazas includes a linear public garden that extends through the commercial center and connects to the regional park system.

FXFOWLE ARCHITECTS, PC

22 West 19 Street
New York, NY 10011
212.627.1700
212.463.8716 (Fax)
info@fxfowle.com
www.fxfowle.com

FXFOWLE ARCHITECTS, PC

FXFOWLE ARCHITECTS, PC

The Helena Apartment Building
New York, New York

The Helena, a 37-story residential building with 600 studio, 1- and 2-bedroom apartments, is the first phase of the firm's master plan for the riverfront block at the Hudson River end of West 57th Street. A composition of interlocking elements — volumes, fenestration, and balconies — produces interesting and varied images of the building, which is visible from many vantage points. The design represents a reinvention of the conventional New York residential building, with floor-to-ceiling glass, wrap-around windows, and sleek metal wall panels. Integrated high-performance technologies such as a blackwater treatment plant, efficient microturbines, and green roofs will earn the building a LEED Gold Certification.

Top left: Glass barrier around planted rooftop.

Middle left: Detail of corner balconies.

Bottom left: End view.

Below: Ground floor plan, with retail space in yellow.

Bottom: Continuous spandrels on long elevation.

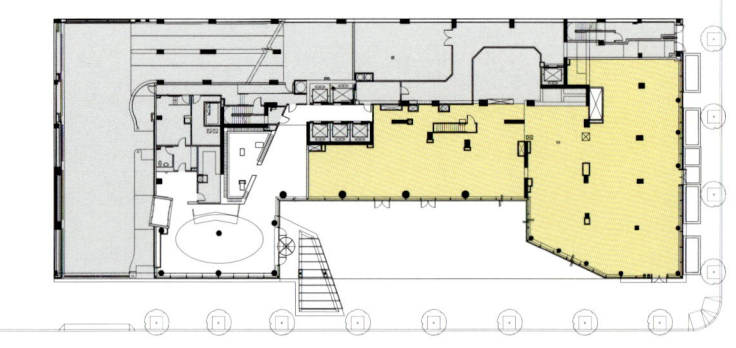

FXFOWLE ARCHITECTS, PC

The New York Times Building
New York, New York

Renzo Piano Building Workshop and FXFOWLE collaborated on the design of the new headquarters for The New York Times. Located on Eighth Avenue between 40th and 41st Streets, the structure will unite the company's employees under one roof in an exceptional signature structure. A grille of glazed terra cotta tubes will screen the floor-to-ceiling glazing. Increased ceiling heights and under-floor ventilation systems will ensure a new standard of comfort and efficiency for high-rise office space. The New York Times will occupy half of the building's 1.6 million square feet, with the remainder leased to retail and corporate tenants by the Forest City Ratner Companies, the project's developer. Construction of the building began in 2005, with completion expected in 2007.

Right, top to bottom: Base of tower and low-rise wing; signature sign across entrance front; exterior wall with terra cotta tube sunscreen; interior view of full-scale office mock-up.

Below: Building in Midtown Manhattan context.

FXFOWLE ARCHITECTS, PC

Whitman School of Management
Syracuse University
Syracuse, New York

The 165,000-square-foot Martin J. Whitman School of Management building is a crucial enhancement of the school's competitiveness. It contains classrooms, offices, and ample space for team meetings and collaborative activity among its 1,400 students, as well as dedicated distance-learning facilities and executive space. The primary design element, a central corridor with a grand staircase at its central lobby, maximizes program connectivity. Some of the many integrated sustainable features include under-floor air displacement, radiant heating and cooling, and highly efficient air filtration and distribution.

Left, top to bottom: Corridor linking facilities at all levels; detail of exterior cladding.

Below: Entry floor plan.

Bottom: Model, showing varied volumes flanking central spine.

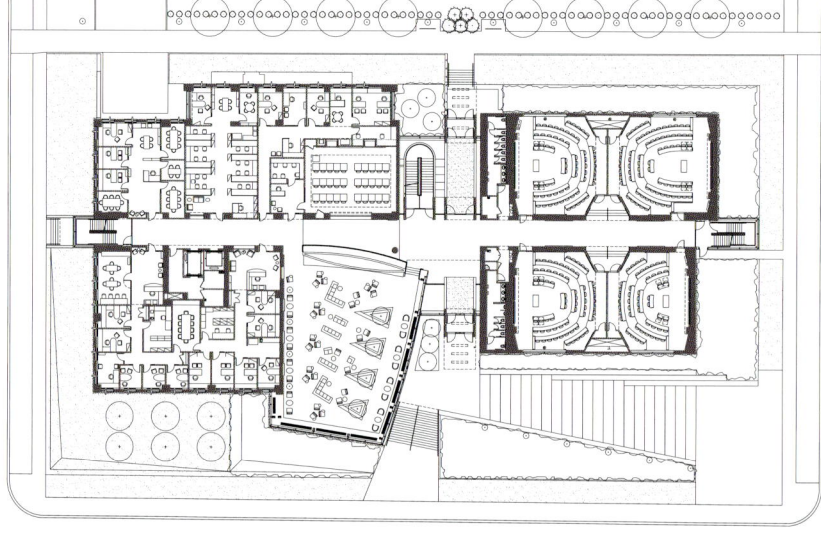

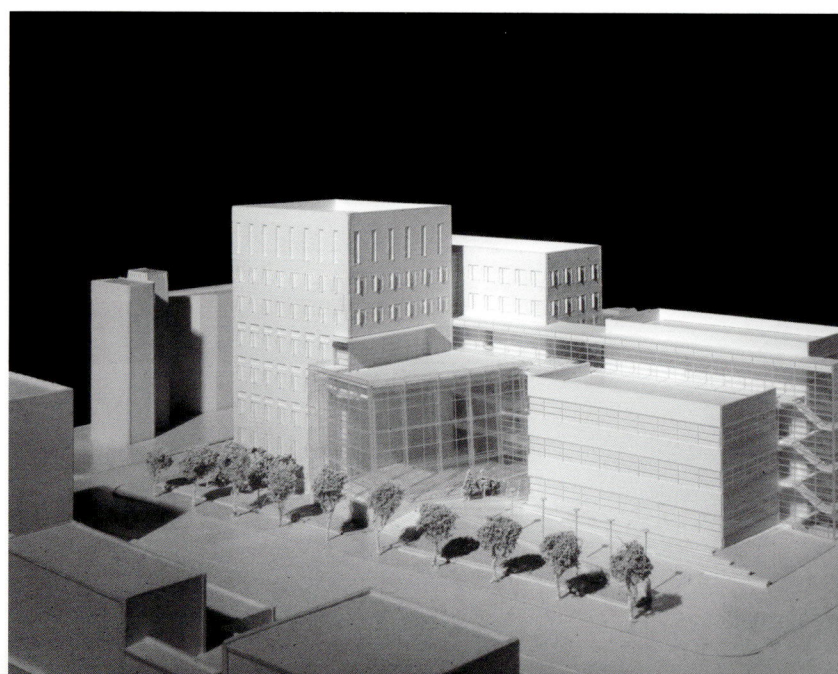

FXFOWLE ARCHITECTS, PC

Lincoln Center Redevelopment
New York, New York

In collaboration with Diller Scofidio + Renfro, FXFOWLE is working to redefine and reactivate the public spaces of the Lincoln Center for the Performing Arts. With the elimination of the massive Milstein pedestrian bridge over 65th Street, the once-forbidding street will become a magnet of activity. A new transparent envelope for the Juilliard building at street level, along with new signage and graphics, will enliven the scene. The extension of the Juillard School and the renovation of the Alice Tully Hall lobbies will expose performing arts activities to both 65th Street and Broadway. Respect for the design character of the original 1960s buildings and public spaces has motivated interventions that will complement the existing while energizing the public experience.

Far left: New Juilliard lobby.
Top: Proposed Juillard Building Broadway front.
Above: New Juilliard School entrance.
Below: Allice Tully Hall after renovations, with 65th Street at left, Broadway at right.

FXFOWLE ARCHITECTS, PC

Tianjin Tower
Tianjin, China

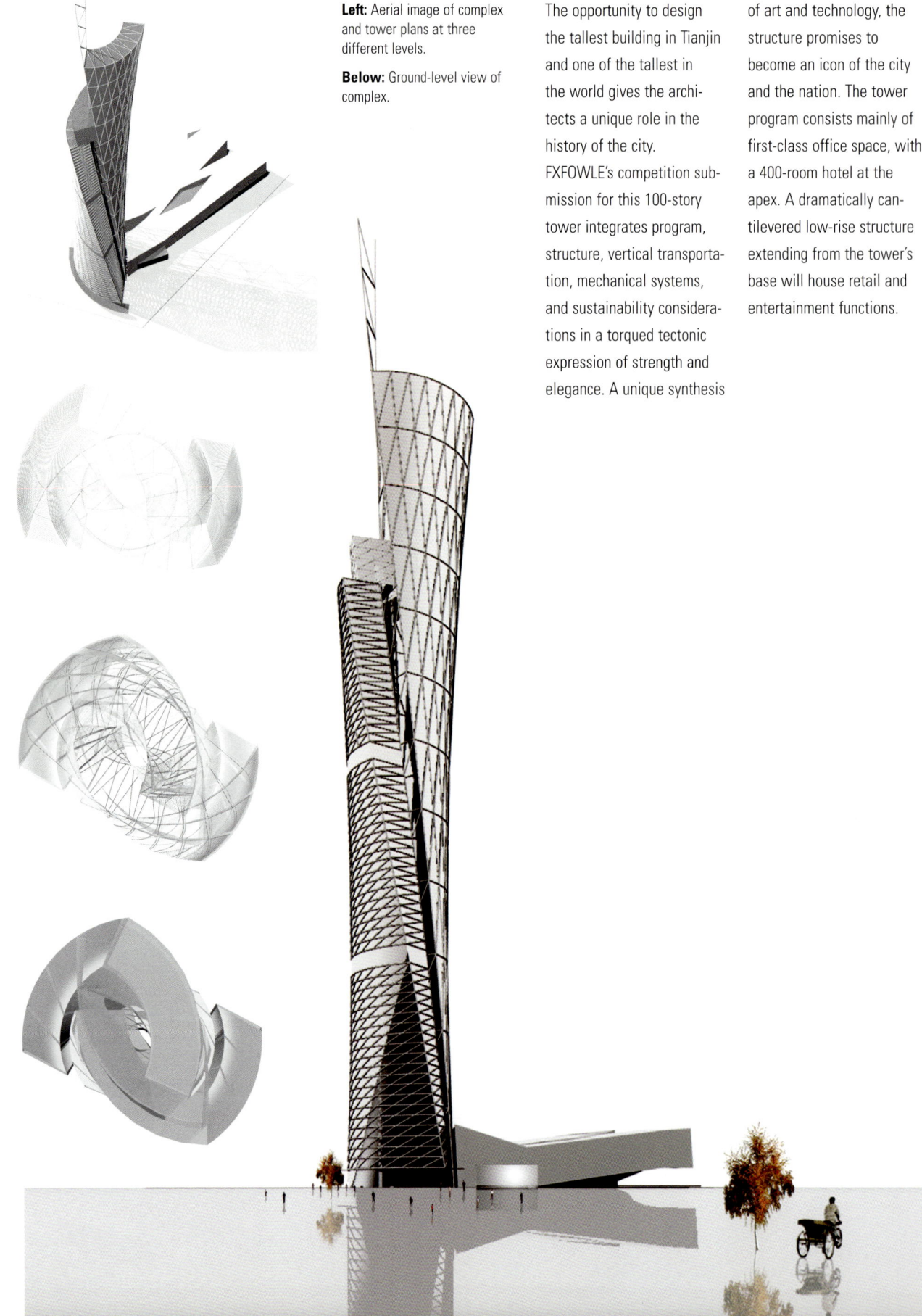

Left: Aerial image of complex and tower plans at three different levels.

Below: Ground-level view of complex.

The opportunity to design the tallest building in Tianjin and one of the tallest in the world gives the architects a unique role in the history of the city. FXFOWLE's competition submission for this 100-story tower integrates program, structure, vertical transportation, mechanical systems, and sustainability considerations in a torqued tectonic expression of strength and elegance. A unique synthesis of art and technology, the structure promises to become an icon of the city and the nation. The tower program consists mainly of first-class office space, with a 400-room hotel at the apex. A dramatically cantilevered low-rise structure extending from the tower's base will house retail and entertainment functions.

FXFOWLE ARCHITECTS, PC

Dosflota Multipurpose Complex Master Plan
Moscow, Russia

Right: Complex seen from riverfront promenade.
Below right: Main tower seen from glazed winter garden.
Below: Skyline view from water.
Bottom: Aerial view of complex.

The planned Dosflota Multipurpose Complex is a unique residential, recreational, and commercial development to be located in the northwestern part of Moscow. The site, once a sports complex, consists of 9.9 hectares (24.5 acres) on the west bank of the Khimkinskoye Vodokhranilishe. The master plan integrates various components to create a resort-like urban center in which hotel, office, conference center, residential, marina, and recreational uses are interconnected with a gracious network of interior and exterior public spaces and amenities.

FXFOWLE ARCHITECTS, PC

Renaissance Place Redevelopment Plan
Naugatuck, Connecticut

FXFOWLE is developing a master plan for the 60-acre Renaissance Place site in the historic downtown of Naugatuck, a once heavily industrial city. With a mix of residential, retail, office, cultural, and educational uses, the plan creates an accessible riverfront, identifiable north and south gateways, and an extension of two main streets as spines of the development. An intermodal transportation center at the train station will connect the site and its 2,000 residential units to regional destinations. The latest sustainability technologies will contribute toward the project's goal of zero net energy use and educate the public on the benefits of energy-efficient design.

Left: Massing of new construction along waterfront.

Below: Diagram of conceptual approach.

Bottom: Aerial rendering of completed development.

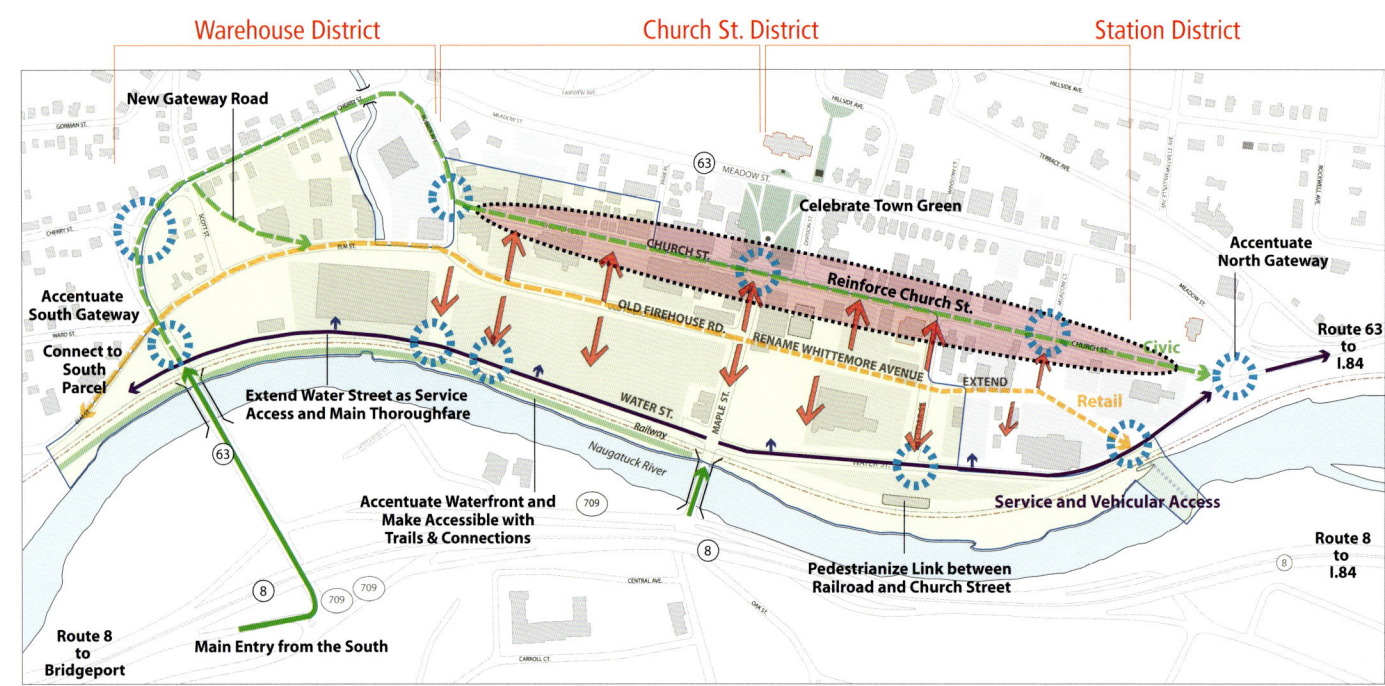

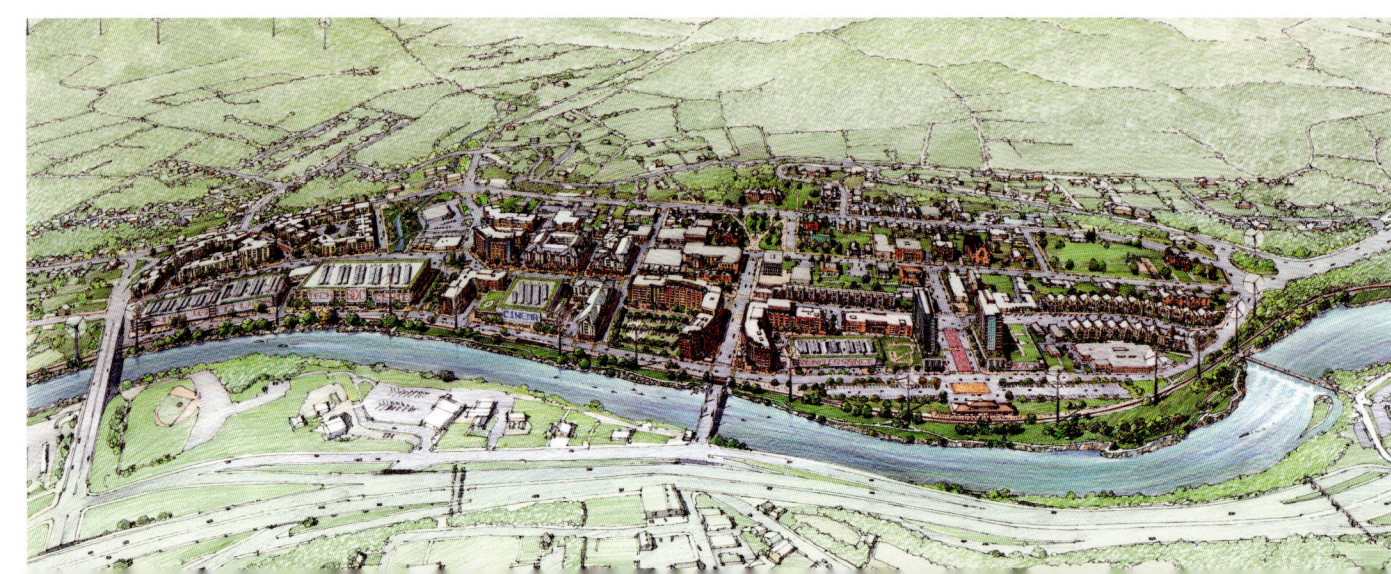

Glatting Jackson Kercher Anglin Lopez Rinehart, Inc.

33 East Pine Street
Orlando, FL 32801
407.843.6552
407.839.1789 (Fax)
corporate@glatting.com
www.glatting.com

1389 Peachtree Street, NE
Suite 310
Atlanta, GA 30309
404.541.6552
404.541.6559 (Fax)

222 Clematis Street
Suite 200
West Palm Beach, FL 33401
561.659.6552
561.833.1790 (Fax)

Glatting Jackson Kercher Anglin Lopez Rinehart, Inc.

Hollis Garden
Lakeland, Florida

Above: The Swan plaza has become a favorite venue for outdoor weddings. The Lakeland skyline is featured beyond.

Left: Appealing plantings draw the interest of children.

Below: Fountain banks and runnels highlight a Wrightian theme and tie the garden together.

Right: Overall view.

Facing page, bottom: Lawn and border; Sobel sculpture.

Photography: Scott Wheeler/City of Lakeland, Glatting Jackson Kercher Anglin Lopez Rinehart, Inc.

Located within Lakeland's historic Lake Mirror Park, this new 1.2-acre garden is a gift to the citizens by the Hollis family. The city contributed in-kind work to the project. The garden's terraces are designed to negotiate the 14-foot drop across the site toward the lake. 16 individual Garden rooms display a spectrum of plantings selected to represent the development of Florida from wilderness through agriculture to formal landscaping. Architectural elements reflect the influences of both Addison Mizner's local interpretations of Spanish vernacular and Frank Lloyd Wright's version of Modernism at nearby Florida Southern College. A focal fountain plaza recalls Michelangelo's geometrical paving at the Campidoglio in Rome. A major work of sculpture installed in the garden is Jeremy Sobel's "Windows."

Glatting Jackson Kercher Anglin Lopez Rinehart, Inc.

Broad Street Park
Baldwin Park, Florida

Below: Central pedestrian bridge over lake, with relocated live oaks at ends and characteristic neo-traditional custom homes beyond.

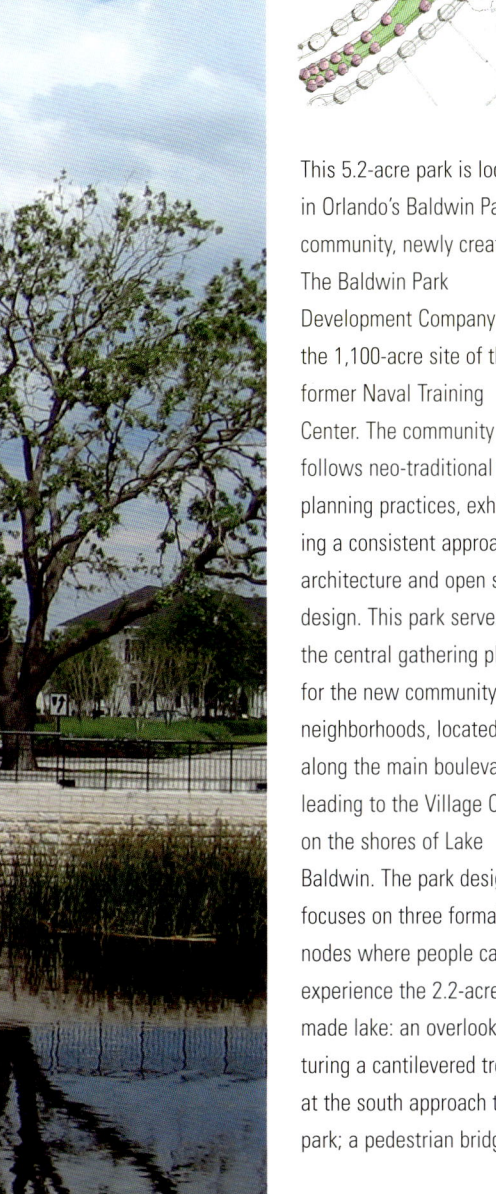

Left: Site plan.

Above: Pedestrian bridge and live oaks.

Below: Cantilevered trellis at south overlook.

Bottom: Perimeter walk approaching bridge.

Photography: Glatting Jackson Kercher Anglin Lopez Rinehart, Inc.

This 5.2-acre park is located in Orlando's Baldwin Park community, newly created by The Baldwin Park Development Company on the 1,100-acre site of the former Naval Training Center. The community plan follows neo-traditional town planning practices, exhibiting a consistent approach to architecture and open space design. This park serves as the central gathering place for the new community's first neighborhoods, located along the main boulevard leading to the Village Center on the shores of Lake Baldwin. The park design focuses on three formal nodes where people can experience the 2.2-acre man-made lake: an overlook featuring a cantilevered trellis at the south approach to the park; a pedestrian bridge at the center that is the primary link between neighborhoods to the east and west; and a formal green at the north end that serves as the principal gathering space. On-site specimen live oaks that were saved and relocated anchor the ends of the bridge and shade its approaches. The formal lawn of the north neighborhood green overlooks the weir fountain that contains the upper source pool and animates the area with the sound of falling water. A walk encircling the park has occasional side paths that provide a more intimate contact with the water's edge. The perimeter path passes through the dense shade of several "cypress domes" that recall the great wild landscapes of Florida.

Glatting Jackson Kercher Anglin Lopez Rinehart, Inc.

Park Avenue Streetscape
Winter Park, Florida

Designed to "right size" the streets through the prestigious Park Avenue shopping district, the master planning and phase I construction of Winter Park's streetscape project was carried out by Glatting Jackson Kercher Anglin Lopez Rinehart, Inc., in association with architects Dover Kohl & Partners. In all, 4,200 feet of pedestrian ways were improved at a cost to the municipality of $4,100,000. Travel lanes were narrowed and sidewalks widened, and traffic calming was effected. Improvements were made to landscaping, paving, site furnishings, and utilities. Specific features include additional live oaks, reused brick street paving, and new standardized trash receptacles. The project has distinctly enhanced the charm and economic health of this unique urban village.

Top left: Accessible crossing.

Middle left: Typical street scene.

Below: Plantings and awnings shading sidewalk.

Facing page top: Sidewalk dining.

Facing page bottom: Scene showing sheltered seating, planters, and streetcorner bump-out.

Photography: Glatting Jackson Kercher Anglin Lopez Rinehart, Inc.

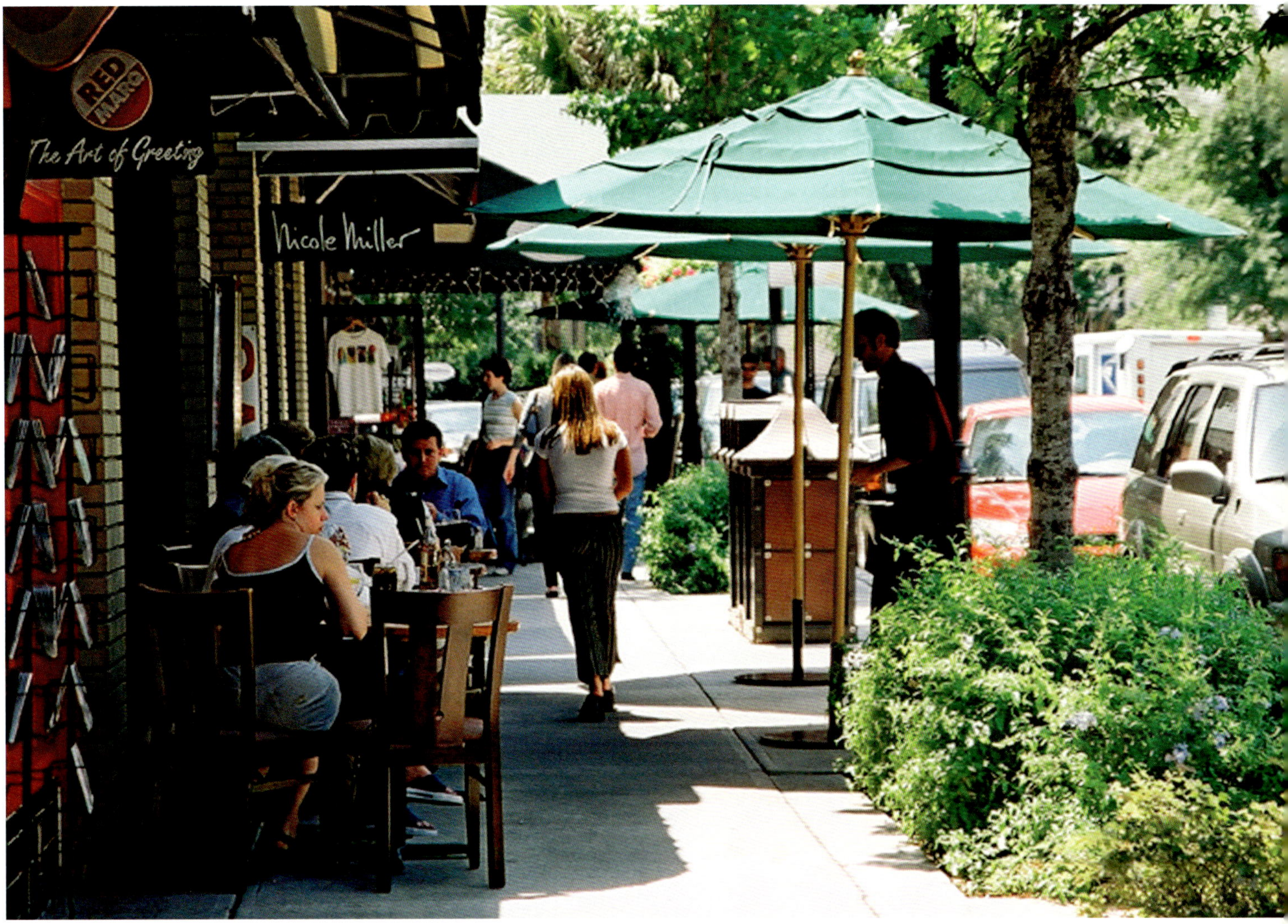
RED MARQ
The Art of Greeting
Nicole Miller

GAP

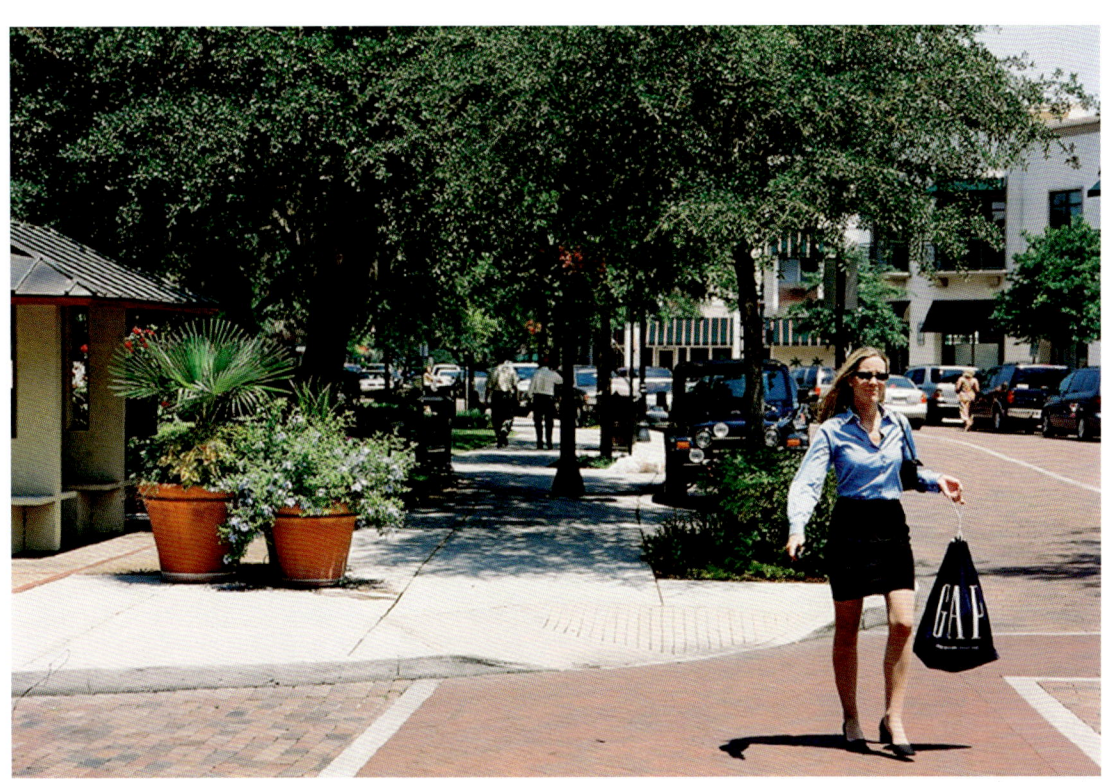

Glatting Jackson Kercher Anglin Lopez Rinehart, Inc.

The Heights
Tampa, Florida

A vibrant mix of commercial and residential development is proposed on this 40-acre, post-industrial site. Through dead-end street and oversized waterfront parcels, the underutilized site has become disconnected from the City of Tampa. The proposed plan will reconnect the area's streets in a grid pattern similar to Downtown Tampa and reopen the district to the water. "Great Streets" will lead to a new Riverwalk, with docks and outdoor cafés, designed to connect to a comprehensive riverfront walk planned by the City. The new townhouses and residential high-rises, which are intended for a diverse population, will enjoy views of the Hillsborough River and the Downtown skyline. The plan features the reuse of the old Tampa Armature Works building as a residential tower including a possible site for a history museum. The plan also revives the existing Waterworks Building as a place for civic functions and a small café. The adjacent Waterworks Park is energized by incorporating it into the Riverwalk and creating places for active and passive recreation within this urban neighborhood.

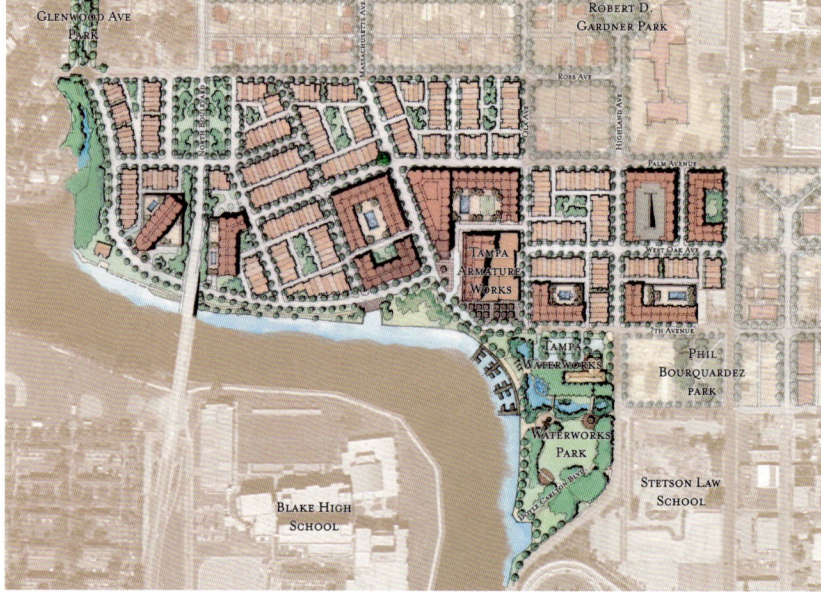

Top left: The Heights in relation to Downtown Tampa.

Above left: The revived Waterworks Building and café.

Left: Riverwalk.

Above: The Heights plan.

Goody Clancy

334 Boylston Street
Boston, MA 02116
617.262.2760
617.262.9512 (Fax)
arch@goodyclancy.com
www.goodyclancy.com

Goody Clancy

Goody Clancy Fort Point Channel
Boston, Massachusetts

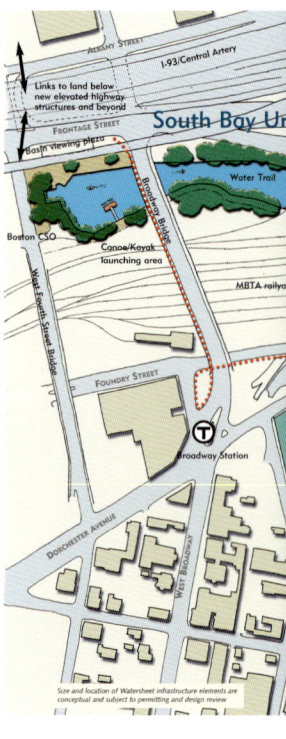

Once a neglected industrial waterway, the Fort Point Channel is being reinvented as a 50-acre water park enlivened by creative uses of its "watersheet." It is being envisioned as a major public destination linking downtown Boston with the South Boston waterfront, the city's newest development frontier. The plan builds on the strengths of existing destinations around the channel, including the Boston Children's Museum, which plans a major expansion of its building there, the Boston Tea Party Ship and Museum, several waterfront restaurants, and New England's largest resident arts community. Between six and eight million square feet of mixed-use development is planned for the edges of the channel over the next several years. Proposed uses of the watersheet include floating classrooms for the Children's Museum, an "art basin" for temporary and permanent works, barges for musical and artistic performances, boat rentals and water rides, model boating, water transportation facilities, transient berthing for educational vessels, and other uses serving residents, downtown workers, and visitors. The plan results from a collaborative effort involving the city, surrounding landowners, environmental advocates, the local arts community, and residents of adjoining neighborhoods. It was jointly

Above: Plan showing proposed features.
Above left: Aerial view looking toward downtown.
Left: Inner "Urban Industrial Wild" portion of channel.

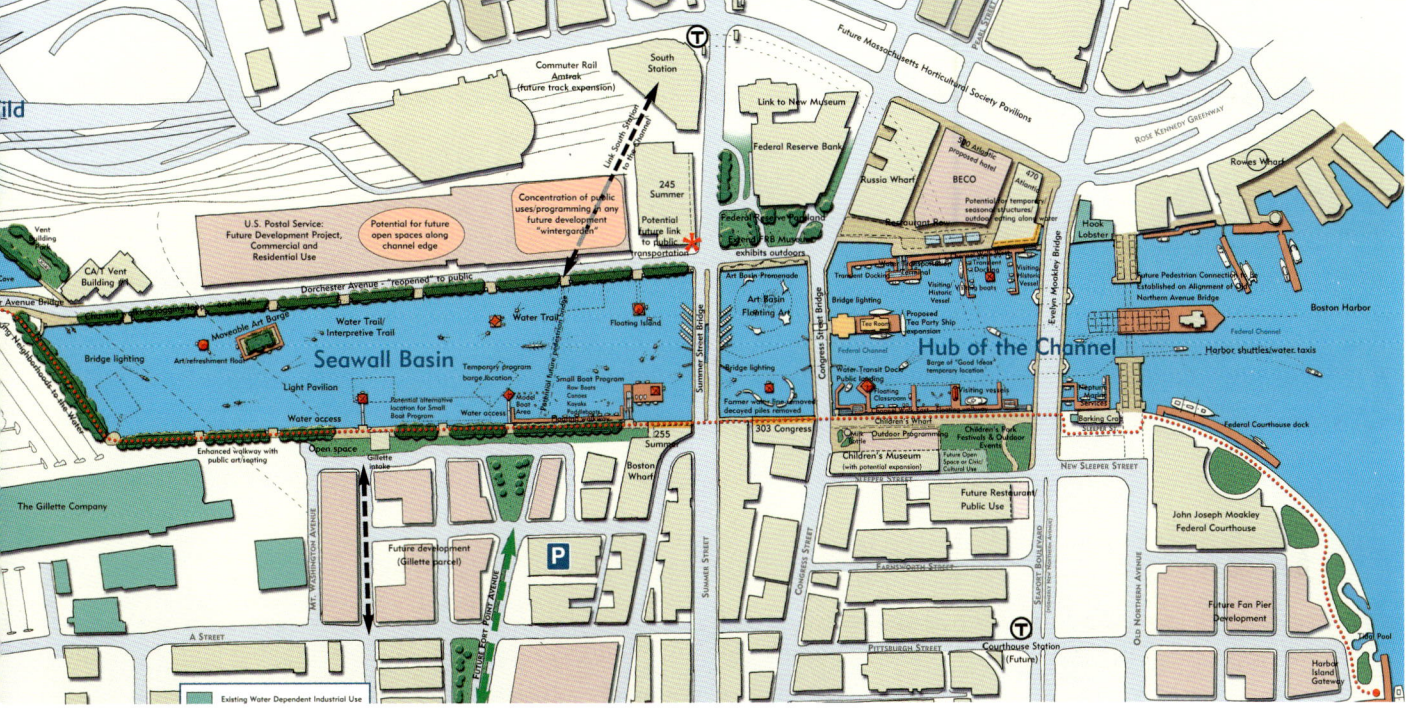

Top left: Completed area of waterfront park.

Top right: Channel as seen from downtown high-rise.

Left: Art installation in channel.

Below left: Potential recreation on "Seawall Basin" part of channel.

187

Left: Northern Avenue Bridge over channel, site of seasonal farmer's market.

Below left: Dining deck outside Boston Children's Museum.

Bottom left: Performance event along channel.

Right: Harborwalk at mouth of channel, with interpretive display.

Below right: Construction view of 500 Atlantic Avenue residential and hotel tower, which will include waterfront restaurant.

Bottom right: Completed portion of park and promenade near Boston Children's Museum.

Photographs: Goody Clancy & Associates.

funded by the city with the Fort Point Channel Abutters, a group that includes the area's museums, major Boston developers, corporations, and real estate interests, and government bodies including the Federal Reserve Bank and the U.S. Postal Service. In 2004, the Friends of the non-profit Fort Point Channel was established to help implement the plan, and it has taken on the long-term programming and maintenance of the waterway and its edges. Water quality improvements are underway, and visitors are already able to enjoy new promenades and open spaces being built in connection with the vast Central Artery/Tunnel project, which adjoins part of the channel. Improved pedestrian connections will be incorporated in the rebuilt Boston Tea Party Ship and Museum, and an additional acre of open space will be provided on the downtown side with the redevelopment of Russia Wharf, 500 Atlantic Avenue residential and hotel project, and Independence Wharf. New waterfront restaurants will be incorporated into 500 Atlantic Avenue and the existing buildings that flank it. Interpretive displays are being placed at key locations to remind visitors of the harbor's maritime history.

Goody Clancy   Assembly Square
Somerville, Massachusetts

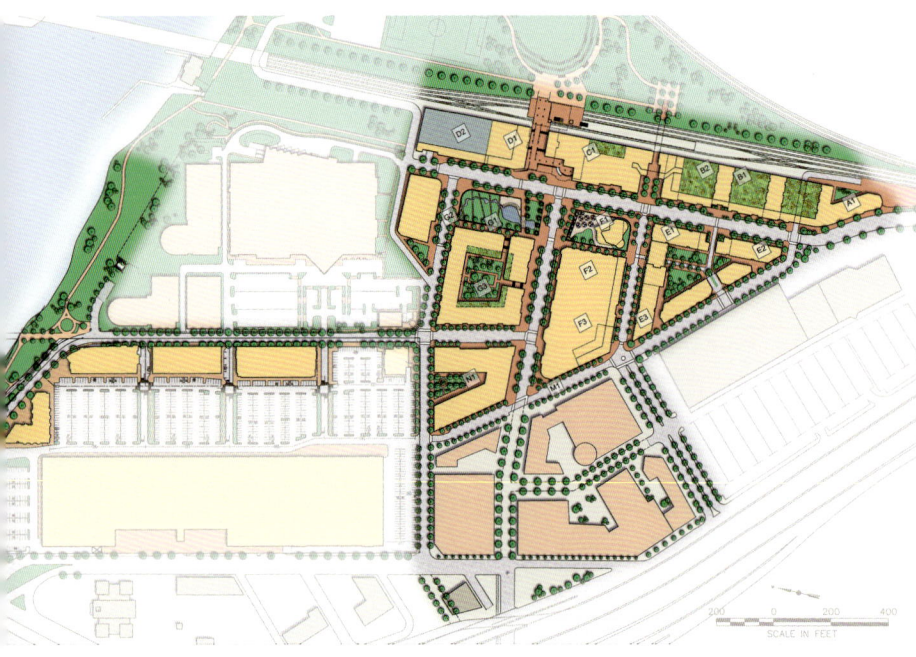

Left: Site plan, showing existing big-box retailers surrounded by mixed-use development.

Below: Phase I conversion of failed strip mall into walkable mixed-use environment.

Facing page, top: Assembly Square, showing transit station and surrounding mixed-use buildings.

Facing page, bottom left: Street-level view of Assembly Square.

Facing page, bottom right: Aerial view of overall development, with Assembly Square at center; view from square toward riverfront.

Renderings: Dongik Lee.

One of the few underdeveloped tracts close to the Boston core, this 140-acre brownfield site has recently attracted big-box retailers. But the current accessibility of the site, combined with its potential for an on-site transit station, argue for mixed-use redevelopment at urban density, which will support a lively public realm lined with street front retail and will justify the costs of site cleanup, the transit station, and other needed improvement. The plan proposes a new urban district, with a strong sense of place shaped around streets and squares, yet successfully integrates existing and proposed big box tenants, locked in place for now by current leases and ownership. Starting with a 26-acre failed retail mall, Phase I will include residential and office development over ground-floor retail, centered on a new pedestrian-oriented "Main Street." Phase II, located on 35 acres of city and privately-owned land, approved by the city of Somerville, will include the Assembly Square public urban space and transit station, with a potential 4 million square feet of development, including 1,300 housing units, over ten years. Eventually, the grid pattern of streets and public squares will be extended over the remaining site.

Goody Clancy North Allston Strategic Framework for Planning
Boston, Massachusetts

In early 2000, Harvard University announced its intention to concentrate its future growth on approximately 100 acres of commercial and industrial land it had acquired in the North Allston section of Boston, across the Charles River from its historic Cambridge campus. The announcement raised concerns that the neighborhood's small-to-medium-sized industrial businesses and established residential areas would be threatened. Recognizing that a move of this scale held major implications for the neighborhood, the city, and the university, the city and North Allston community members agreed to collaborate with Harvard on a four-year planning process, involving many interviews, workshops, public meetings, and other sources of input to produce this Strategic Framework. Setting forth principles for housing, open space, transportation, and economic development, the Framework creates a model of growth, change, and preservation to reshape a newly blended academic, residential, commercial, research, and recreational neighborhood; it embodies a desire, agreed to by all stakeholders, for full integration of "town" and "gown."

Above: Plan with parts of existing Harvard campus along river at upper right.

Below left: Proposed development near center of plan.

Below right: Mixed institutional and residential district.

Bottom left: Presenting the Framework to the public

Bottom right: Mixed-use development along major avenues.

Lessard Group Inc.

8521 Leesburg Pike
Suite 700
Vienna, VA 22182
703.760.9344
703.760.9328 (Fax)
info@lessardgroup.com
www.lessardgroup.com

Lessard Group Inc.

Trump Plaza
New Rochelle, New York

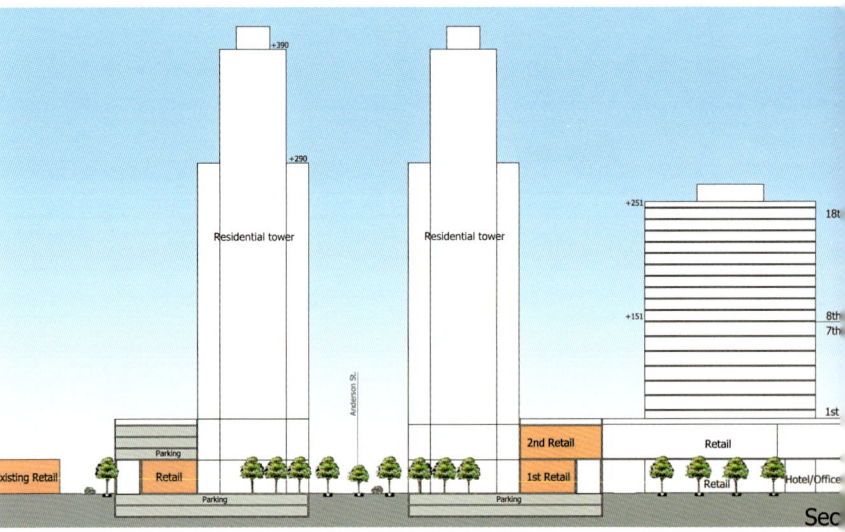

Above left: Long-range plan, with Phase I in upper right quadrant.

Above: Schematic elevation, with Phase I at right, matching Phase II tower at left.

Left: Curved entrance volume at key intersection.

Situated on a 1.9-acre site in the center of downtown New Rochelle, this 320,000-square-foot mixed-use structure is a key component of the city's plan for revitalization and redevelopment. The design was inspired by the idea of creating a signature building in the heart of New Rochelle, a growing market just outside New York City. The project features a 32-story high-rise of about 149,600 square feet, containing 181 residential units, with retail on the first two floors and 240 underground parking spaces for residents' use. Parking for the retail is provided in an adjacent municipal garage, reached via a pedestrian bridge. The massing and detail of the buildings complement and enhance the existing commercial district. Phase I of development is scheduled for completion in 2007. Future phases are planned to roughly quadruple the initial square-footage, with a varying mix of residential, office, retail, and parking facilities, and will include a twin to the Phase I residential tower, the two flanking a landscaped public mall.

Above: Main entrance to tower containing apartments above retail base, with pedestrian bridge to retail parking in existing public garage.

Right: Perspective and elevation drawings of mixed-use tower.

Lessard Group Inc.

Springfield Town Center
Springfield, Virginia

Located at the intersection of I-95 and the Capital Beltway, this mixed-use urban center is well situated to revive and expand the retail and residential resources of Springfield. The master plan of the 8.86-acre site is organized to reduce the impact of noise on the residential component. A parking structure and a hotel form a continuous barrier to buffer the noise of the highways. Internally, public plazas and retail along the streets will help make the development pedestrian-friendly. The completed project's residential component of 912,715 square feet will provide the consumers that are vital to pedestrian-oriented retail. The total 1,157,715-square-foot program includes 100,000 square feet of retail, 100,000 square feet of hotel, and 40,000 square feet of offices, plus 2,044 required parking spaces. An allocation of 5,000 square feet of interior community space, along with extensive public open space, will enhance the development as a center for the existing community.

Opposite top: Aerial view of completed development, with highway interchange at lower left.

Opposite middle: Master plan, with hotel and parking structure buffering internal residential buildings, plazas, and pedestrian-friendly streets from highway noise.

Opposite bottom: Streets at core of development.

Above: Mixed retail, office, and residential buildings along curving street.

Below: Plaza and skylighted arcade between residential high-rises with mixed-use lower floors.

Lessard Group Inc.

Canton Crossing
Baltimore, Maryland

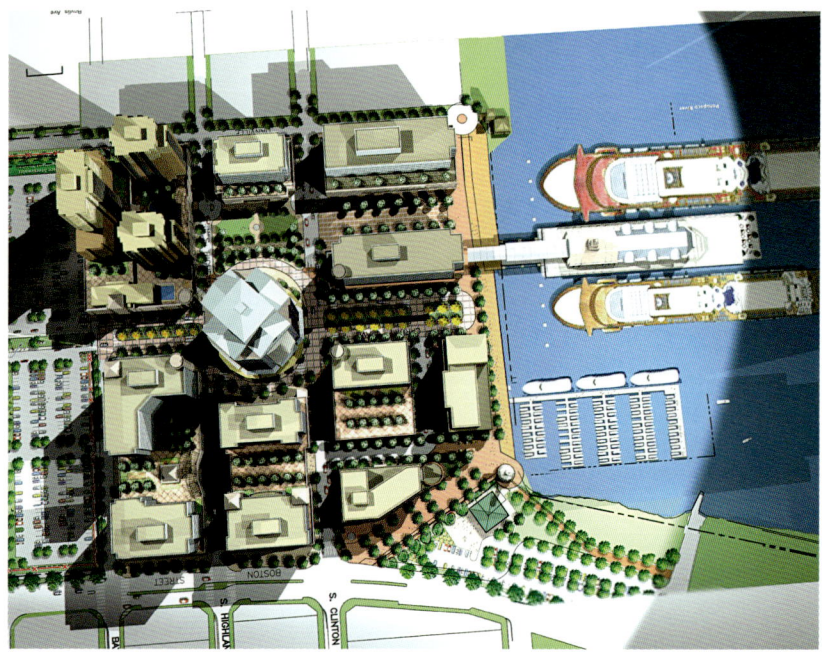

Situated along the edge of the Patapsco River, this 32-acre waterfront development introduces a mixed-use urban cityscape to an existing predominantly industrial location. The development features a variety of high-rise buildings that include multi-family, commercial office, retail, and hotel components. A central polygonal office tower marks the project's highest point. The massing and heights of surrounding buildings respect a hierarchy of forms. The building masses typically sit atop four-story parking garages that are masked at ground level by retail and restaurants. View corridors extending into the site from the waterfront allow for optimum views from most buildings. Streets are activated with wide sidewalks that serve as outdoor sitting areas. The waterfront edge is energized by a boardwalk lined with restaurants. A cruise ship terminal and customs station share the waterfront with a marina that is proposed to have a luxury residential component.

Opposite top: Development as seen from Patapsco River.

Opposite bottom: Aerial view of developed site.

Left: Residential tower complex at lower right in master plan.

Below: Master plan.

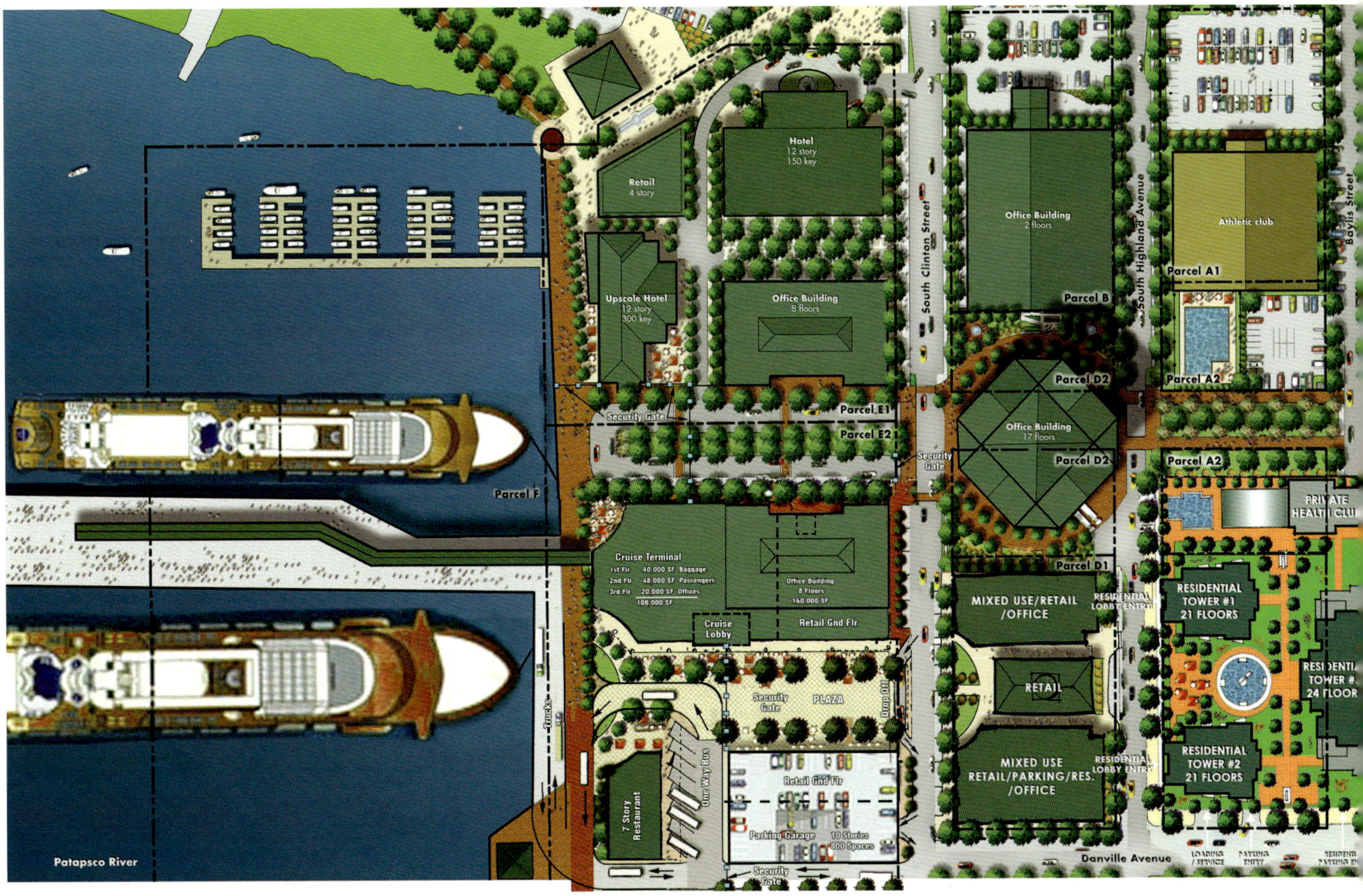

Lessard Group Inc.

National Harbor
National Harbor, Maryland

This mixed-use development will serve as a gateway project for Prince George's County, introducing an urban lifestyle to the area's existing suburban fabric. The site along the Potomac River offers impressive views across to Alexandria, Virginia, and up-river to Washington, D.C., and the sloping terrain facilitates view corridors toward the water. The overall master plan proposes 2,500 residential units in a mix of high-rise, mid-rise and low-rise structures, which are supported by substantial retail, restaurant, and office components. A convention center and a hotel complete the urban mix. The plan divides the 130-acre site into three distinct areas, known as Downtown, Midtown, and Uplands, each identified with a lifestyle and a market segment — with commercial uses concentrated in the Downtown, luxury residential in Midtown, and active adult units in the Uplands. A dedicated tree preserve at the center of the development acts as a central park. The waterfront is activated by restaurants, a theater, and public amenities such as a ferry terminal.

Right: Conceptual master plan.

Below: Uplands residential buildings rising beyond central park and clubhouse.

Bottom: Elevation of a Downtown mixed-use block.

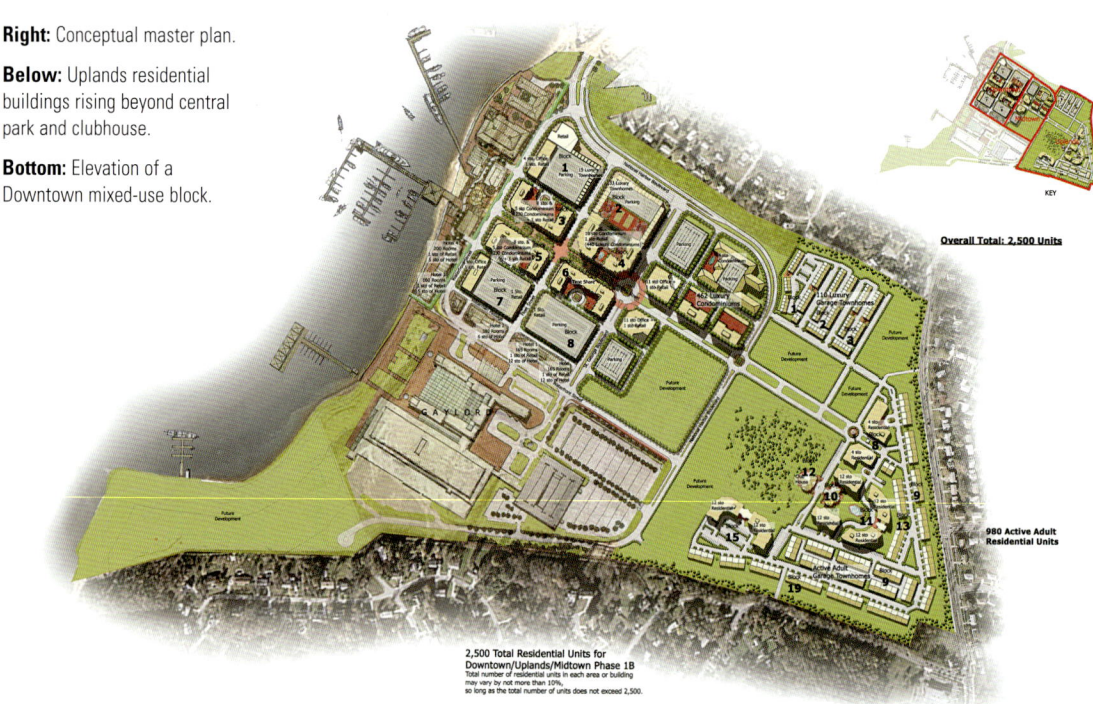

Looney Ricks Kiss

175 Toyota Plaza
Suite 600
Memphis, TN 38103
901.521.1440
901.525.2760 (Fax)

info@lrk.com
www.lrk.com

209 10th Avenue South
Suite 408
Nashville, TN 37203
615.726.1110
615.726.1112 (Fax)

182 Nassau Street
Suite 201
Princeton, NJ 08542
609.683.3600
609.683.0054 (Fax)

31 Main Street
Rosemary Beach, FL 32461
850.231.6833
850.231.6838 (Fax)

671 Front Street
Suite 220
Celebration, FL 34747
407.566.2575
407.566.2576 (Fax)

Looney Ricks Kiss

Looney Ricks Kiss

Jefferson at Providence Place
Providence, Rhode Island

The first luxury housing development in downtown Providence in over 20 years, this 330-unit project is meant to fill the unmet needs of young professionals. At the same time, it is designed to respect the character of historic brick warehouses in its district. The detailing of the facades, with broad arches, storefronts, bay windows, awnings, and individual stoop entries to some street-floor units supports a traditional urban atmosphere. Land costs in this location dictated the density of 75 units per acre, with garage parking. The four-story parking structure aligns at each floor with the

double-loaded corridors of the apartment building. The existing street has been improved with new sidewalks, landscaping, and street lighting and provides a direct pedestrian connection to the city core. The apartments follow 14 different floor plans, ranging in size from 563 to 1,421 square feet. Most of the top-floor units have "loft" studio layouts, with upstairs bedrooms overlooking two-story volumes. Shared amenities include an outdoor pool, a spa, a fitness center, a coffee bar, concierge services, a business center, and a well-equipped club room available for resident functions. Construction on land contaminated with mercury imposed several restrictions. This new construction had to accommodate itself to factory foundations and footings that could not be removed because of contamination. Landscaped areas that could not be capped with impervious construction had to be topped with 18 inches of clean fill.

Facing page: Street views of residential buildings.
Above: Club room.
Right: Interface with city.
Below: Lobby.
Photography: Frank Giuliani.

Looney Ricks Kiss

FedExForum
Memphis, Tennessee

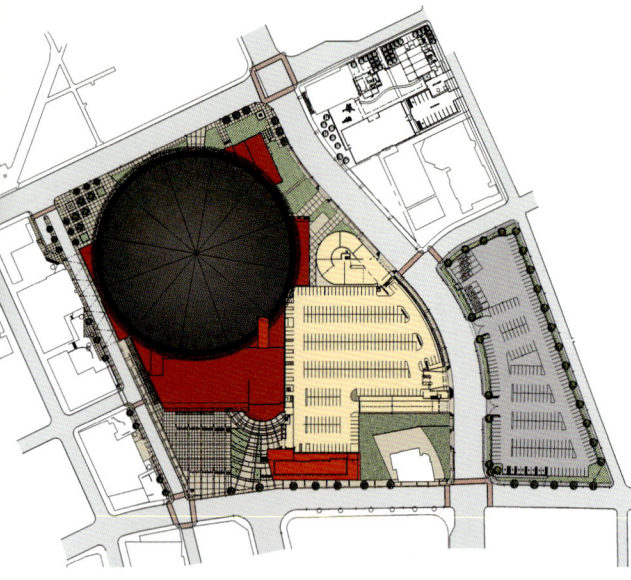

The revival of downtown Memphis has been given a major boost by this $200-million sports and entertainment complex, the largest public building project in the history of Memphis. It is home to the Memphis Grizzlies NBA team plus the Smithsonian Institution's Rock 'n' Soul Museum. The arena seats 18,200 on five levels, with the team's offices on the upper floors of the adjoining four-story office building and the museum on its street floor. Sinuous rerouting of a major street allowed for a 1,500-car parking structure adjoining the arena, which helps fill parking needs for the adjacent Beale Street musical entertainment district. Outdoor entertaining and party areas are accommodated on 35,000 square feet of plaza. Designed by Ellerbe Becket in association with Looney Ricks Kiss, the complex has been adjusted to the scale of historic Beale Street and the surrounding area. The height of the arena was lowered by depressing the event floor 33 feet below street level, simultaneously improving circulation by placing seating areas both

above and below entry level. Also minimizing the impact of the project are the smaller annexes surrounding the central drum, which create the impression of incremental development over a period of time. The arena drum is clad in aluminum panels, its lobby largely in glass, and the smaller structures in a variety of traditional brick. Artwork throughout the development represents the Memphis area's rich cultural and musical heritage. The complex was completed in 2004 on time and within budget.

Facing page: Site plan and aerial view.

Above: View toward entry, showing low wings in scale with surroundings.

Below: Entry plaza.

Photography: Timothy Hursley/The Arkansas Office.

Looney Ricks Kiss

Ave Maria Town Center
Collier County, Florida

The new town of Ave Maria is proposed for 633 acres of land adjoining the developing new campus of Ave Maria University. A prominent oratory, designed by others, is to be the main symbol of the university, at the same time linking it to the town. Looney Ricks Kiss's contribution is the planning of the town center, embracing and radiating from the Ellipse Road around the oratory. Town center planning calls for buildings that complement the central landmark, generate intimate open and covered spaces, and reflect regional traditions. LRK is designing four of the initial six mixed-use buildings in the town core, accommodating 77,800 square feet of retail/commercial space and 70,000 square feet of residential. Construction is to start in 2005, with completion in 2006.

Left: Plan of town center.

Below left: Town center street scene.

Below: Mixed-use buildings along Ellipse Road.

Bottom: Streetscape elevations.

Looney Ricks Kiss

Ross Bridge Village Center
Birmingham, Alabama

The new American resort town of Ross Bridge is anchored by a new 400-acre Town Center, located along the edge of a scenic parkway. The Town Center is entered through its main green and is composed of 75,000 square feet of commercial/retail space, 108 higher-density single-family residential units — cottages and townhouses — and 25 live-work type buildings. The town's architectural vocabulary is defined by that of local and regional villages of the early 1900s. This is evident in the community's town hall and welcome center, patterned after a vintage railroad station. Paired parks of 80,000 and 86,000 square feet will provide open space at the entrance to Ross Bridge, with the Town Center serving as a transition to the surrounding residential areas.

Top: Town Center plan.
Top right: Town aerial view.
Middle right: Market Street buildings.
Above: Town Center residential street.
Right: Main street into town.

Looney Ricks Kiss

Thornton Park
Orlando, Florida

This 809,000-square-foot mixed-use building with 311 condo units serves as a connector between downtown Orlando and the historic residential neighborhoods that line Lake Eola and its parks. The building's five-story mixed-use podium harmonizes with the surrounding low- to mid-rise neighborhood, and the residential tower rising from its corner joins the downtown skyline. A private club, with pool, and rooftop villas are located on the fifth-floor setback. The ground-floor retail was designed to comply with the city's mixed-use ordinance developed for the district. The main retail occupant, a food market, is given a prominent corner presence, with much of its bulk behind the streetscape of smaller retail spaces. By locating the garage entrances on opposite streets, access to underground parking for the retail has been clearly separated from the motor court entry and resident parking.

Above left: Condominium rising at intersection of Central and Lake Avenues.

Left: Corner Grocery entry.

MBH Architects

1115 Atlantic Avenue
Alameda, CA 94501
510.865.8663
510.865.1611 (Fax)

1300 Dove Street, Suite 100
Newport Beach, CA 92660
949.757.3240
949.757.3290 (Fax)

www.mbharch.com

MBH Architects

West Hollywood Gateway
West Hollywood, California

Above: Project's signature night-lighted curves.

Above right: Identifying lighted tower above plaza.

Left: Outdoor dining on widened sidewalks around complex.

Opposite bottom: Plaza with escalators connecting two retail levels.

Photography: RMA Photography.

This complex sets a valuable precedent for integrating big box retailers into an urban, pedestrian-oriented setting. Target and Best Buy have leased space in the development which includes 254,000 square feet of retail on two levels, with space for additional retailers, restaurants, and a 70,000-square-foot community meeting area, all organized around a landscaped public plaza. Widened sidewalks with space for outdoor dining add further to the public realm, and two levels of parking are provided under the complex. The historic Formosa Café is retained and integrated into the design of the project. Both retailers have experienced exceptional sales in this unaccustomed setting. Located at a highly visible intersection, the project has already inspired upgrading of nearby properties. The 4.84-acre site came with several liabilities: multiple ownership, soil contaminants from a former rug-cleaning plant, and existing hard-to-relocate occupants (a car wash, an auto repair shop, and a cast stone manufacturer). Working with city and federal authorities, the development team was able to get a grant and a favorable loan through HUD and to acquire some property through eminent domain.

MBH Architects

200 Brannan
San Francisco, California

This 191-unit residential complex, located in the historic South Beach neighborhood, has an exterior composition of red brick, white plaster, and metal-framed glazing that blends well with the area's warehouse tradition. Facing inward toward a lush central court, 79 of the units have transparent envelopes of sheer glazing, two-stories for each unit, forming a serrated boundary that expresses the individual residential module. Thirteen penthouse units have private balconies with sweeping views. Structural systems are pulled back from exterior walls to leave views unobstructed while meeting stringent seismic requirements.

Left: Interior court, with serrated edge of glazed walls.

Opposite left: Glazed walls revealing two-story unit heights.

Opposite top right: Street elevation, with areas of brick and plaster.

Opposite top center: Elegant complex identification signage.

Opposite bottom: Façade view from street.

Photography: Farshid Assassi.

MBH Architects

The Town Center at Levis Commons
Perrysburg (Toledo), Ohio

The Town Center is the first phase of J. Preston Levis Commons, a 200-acre urban village that will include offices, residential buildings, and recreational areas. This 319,000-square-foot mixed-use center is the focus of a master-planned street grid. The center is anchored by a twelve-screen cinema. Specialty retail, a variety of restaurants, 70,000 square feet of upper level offices, and entertainment venues overlook a town green that includes fountain, benches, and other amenities. Architectural design elements reflect the scale and tradition of the region, with varied building envelopes inspired by Georgian, Federal, Italianate, and Queen Anne precedents.

Opposite top: Building facades of varying form in consistent muted color.

Opposite bottom left: Central fountain on town green.

Opposite bottom right: View of Town Center from Archway.

Above: Town green and flanking buildings, showing on-street parking.

Right: Fountain and retail row with corner turret.

Photography: Tom Ethington.

MBH Architects

Marina University Villages
Marina (Fort Ord), California

Developed adjacent to California's beautiful and historic coastline highway, Marina University Villages is the evolution of the regional lifestyle center. Totaling 700,000 square feet the pro-project contains a collection of restaurants and entertainment venues, specialty and large retailers, along with area suitable for civic events and public art. Marina University Villages is the anchor and first phase of an overall 270-acre master-planned community that ultimately will include single- and multi-family housing, additional neighborhood retail and office space connected with pedestrian and bike friendly streets, open space linkages, transit corridors and vistas to civic landmarks.

Top: Activity in mixed-use town center.

Far left: Town center's mix of automobile and pedestrian traffic.

Left: Portion of town center, showing buildings of various heights.

Below: Elevation of town center.

McLarand Vasquez Emsiek & Partners

1900 Main Street
Suite 800
Irvine, CA 92614
949.809.3388
949.809.3399 (Fax)
info@mve-architects.com
www.mve-architects.com

350 Frank H. Ogawa Plaza
Suite 100
Oakland, CA 94612
510.267.3188
510.267.3199 (Fax)

McLarand Vasquez Emsiek & Partners

McLarand Vasquez Emsiek & Partners

Fruitvale Village
Oakland, California

A number of smart-growth design and land-use concepts are embodied in this transit-oriented, high-density, mixed-use project, which is intended to revitalize Fruitvale's business district. The 10-acre development brings to life an existing BART station by replacing the on-grade parking lot with a commercial, retail, and entertainment paseo that accommodates community-related uses. Capitalizing on the pedestrian traffic generated by the BART station, Fruitvale's redevelopment plan includes over 30,000 square feet of retail and restaurant space, 60,000 square feet of offices, a 40,000-square-foot health clinic, a 12,000-square-foot community resources center, 5,000 square feet library, and 47 live/work units. The two buildings house retail on the first level, community facilities on the second level, and innovative loft housing on the third level. The atmosphere of a lively Mediterranean village is recalled in the project's use of simplified traditional architectural forms, a variety of warm colors, palm trees, and fountains. Circulation on third-level terraces, reached by broad stairs, adds a third dimension of activity and visual interest.

Above: Identifying signage at one entrance.

Left: Site plan.

Above: Plaza approach to BART station.

Top right: Tiered building defining plaza.

Far right: Observable playground.

Right: Library

Photography: Keith Baker Photography; Evelyn Johnson Photography (opposite left).

McLarand Vasquez Emsiek & Partners

The Promenade at Rio Vista
San Diego, California

The Promenade is a 13.8-acre mixed-use community that exploits the site's light-rail transit and riverfront trails. It consists of six structures of four stories over a two-level subterranean garage, with commercial and retail activities surrounding a commons. The commons serves as the heart of the community for residents and transit riders alike. The space is enlivened by arcades fronting on the commercial uses, with patios for outdoor dining and a large fountain as a memory point. Based on the architecture of Irving Gill, the light-colored rectangular forms of the buildings, pierced by round arches, acknowledge the benign climate and design traditions of the area. Trellises, pergolas, courtyards, and patios tie the interiors to the landscape. A network of private streets and pedestrian passages connects destinations inside and outside the project. Retail parking is parallel to the streets or integrated into guest parking of the garage, with access to the commons through the building lobbies.

Above: Commons, with axial tower marking trolley station.

Below: Central fountain.

Right: Commons by day, showing plantings and parallel retail parking.

Photography: Robert A. Hansen Photography.

McLarand Vasquez Emsiek & Partners

Hollywood & Vine
Hollywood, California

Above: Alternative Vine Street treatments.

Below: View with Vine Street at left.

With one of the most famous intersections in the world as its address, this mixed-use development will deliver a Hollywood experience. One of Los Angeles' many MTA transit villages, the project will incorporate an existing subway terminal. The development responds to historical Hollywood with the use of colorful materials and established cornice heights. It also expresses the present and future of Hollywood through its incorporation of large-scaled wall graphics. In addition to MVE & Partners, the project design team includes HKS Architects for the 300-room hotel and the 140 condominiums. A transit plaza located on Hollywood Boulevard at the key subway entrance is lined with shops and restaurants and provides a forecourt for the hotel and condominium buildings. Vine Street is the primary address for the 350 apartments, designed in an urban loft aesthetic, which will range from 550 to 1,150 square feet each. Below these will be neighborhood retail space at street level.

McLarand Vasquez Emsiek & Partners

Uptown
Oakland, California

Located just north of the recently renovated Oakland Civic Center, this 14-acre project will transform underutilized parking lots into a community designed to attract residents downtown. It will include 60,000 square feet of retail/commercial space, 1,200 mixed-income apartments, 400 off-campus graduate student housing units for U.C. Berkeley, and 900 condominiums. The corridor along Telegraph between the landmark Paramount Theater and the future Fox Theater renovation will accommodate live events, a food court, and a revitalized existing skating rink. Low-rise rental units and loft-style apartments, intended for young professionals, will be linked to the entertainment district by pedestrian paseos. The high-rise towers along Telegraph will house for-sale units. Loftlike units will be created by renovating existing structures, thereby maintaining the historical character of the site.

Top: Birds-eye view of site.

Above: Renovated Fox theater on Telegraph entertainment corridor.

Left: Park at center of development.

McLarand Vasquez Emsiek & Partners

Douglas Park
Long Beach, California

MVE & Partners worked with Boeing Realty Corporation in master planning their 260-acre former aircraft factory property adjacent to Long Beach Airport. The production site for Douglas Aircraft during World War II, the factory later turned out McDonnell Douglas jet-liners. As the reuse of this tract was of great interest to the community, MVE participated with Boeing in a series of public workshops soliciting comments and support from stake-holders, while also working closely with city officials and staff. The master plan features 3.3 million square feet of job-generating commercial uses ranging from offices and research and development to light industry and distribution. A 100-acre residential district with about 1,400 dwelling units will include a variety of detached and attached types along tree-lined streets. A mixed-use district with key retail amenities will serve as a focal point, as will numerous parks and open spaces.

Top right: Conceptual land use plan.
Below: Aerial perspective of entire development.
Middle right: Neighborhood park.
Bottom right: Residential street.

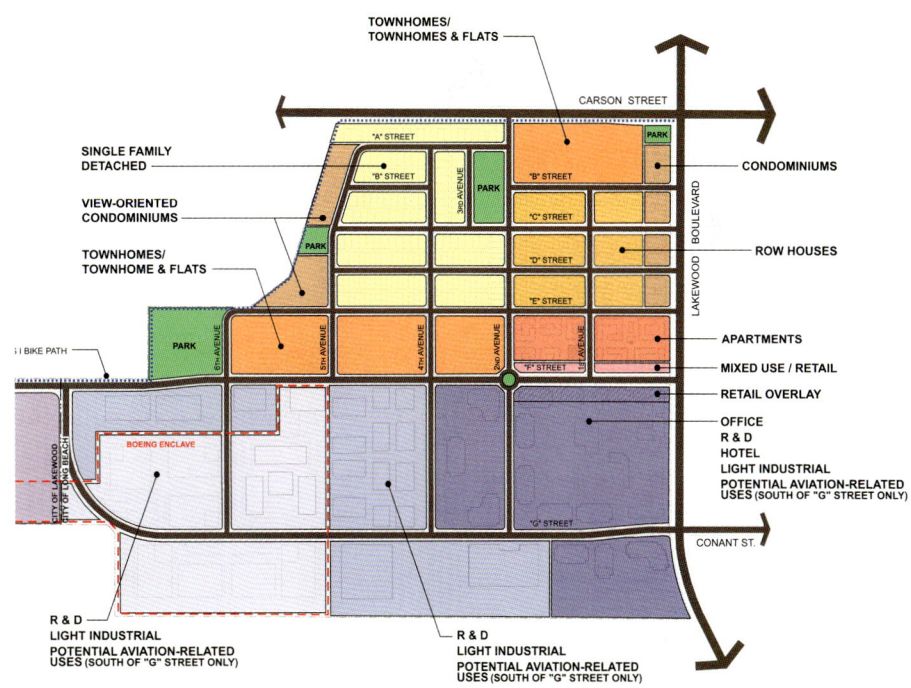

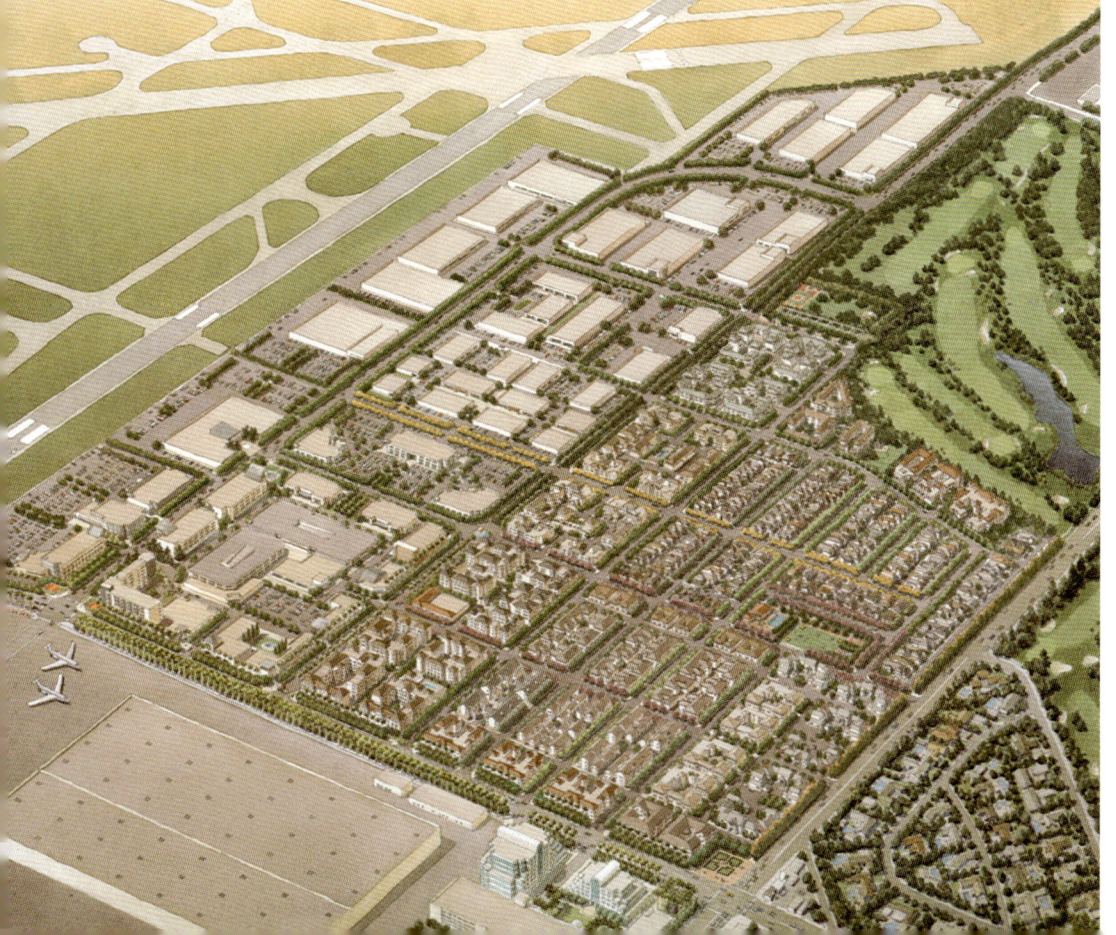

McLarand Vasquez Emsiek & Partners

Tralee
Dublin, California

Left: Village plan, with mixed-use retail center at lower left.

Below left: Residential street.

Bottom: Mixed-use activities and on-street parking in Village Center.

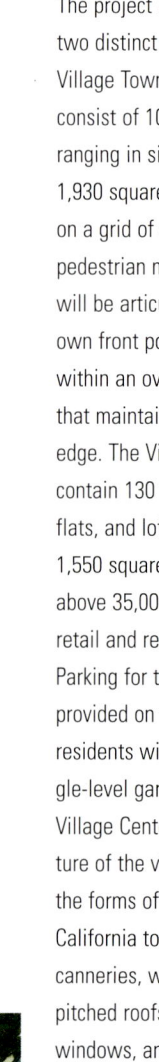

A mixed-use urban village will be developed on the ten-acre site of a vacant retail center, adjacent to Dublin's civic center, sports park, and business district. The project is composed of two distinct parts. The Village Townhomes area will consist of 103 rowhouses, ranging in size from 1,350 to 1,930 square feet, organized on a grid of streets and pedestrian mews. Each unit will be articulated, with its own front porch or stoop, within an overall envelope that maintains the street edge. The Village Center will contain 130 townhouses, flats, and lofts, from 600 to 1,550 square feet in size, above 35,000 square feet of retail and restaurant space. Parking for the retail will be provided on the streets, and residents will park in a single-level garage below the Village Center. The architecture of the village evokes the forms of historic California towns, farms, and canneries, with vaulted and pitched roofs, numerous bay windows, and generous use of brick. Metal canopies will provide shelter and incorporate signage, as they have traditionally in local towns.

Pappageorge/Haymes Ltd.

814 North Franklin
Suite 400
Chicago, IL 60610
312.337.3344
312.337.8009 (Fax)
email@pappageorgehaymes.com
www.pappageorgehaymes.com

Pappageorge/Haymes Ltd.

The Glen Town Center
Glenview, Illinois

Left: Aerial perspective, with control tower at center.

Right: Mixed-use row, with broad sidewalks and angled parking.

Below: Navy Park and mixed-use façade.

Photography: Pappageorge/Haymes Ltd.

As the high-density, mixed-use focus for the 1,100-acre redevelopment of a former naval air station, Glen Town Center is designed to evoke the classic shopping districts of nearby Lake Forest and Winnetka. On its 45 acres, it includes 110,000 square feet of retail, restaurants, and movie theater, 181 apartments above street-floor commercial, and 154 townhouses. Two groups of gabled townhouses are located at either end of the center, curving around quiet greenswards, and others are "laminated" onto parking garages to screen them from view. The main retail street follows a graceful arc, linked at either end to the area's main boulevard. Centered on this arc and set off by the new Navy Park are the restored 1930s air control tower and hangar.

Facing page middle: One of residential ovals; reused air station building facing Navy Park.

Facing page, bottom: Axial entry to center, aligned with control tower.

Unifying the development is a vocabulary of red brick walls, slate roofs, punched residential windows, and projecting bays and balconies. Once established, this vocabulary has been varied by irregularities in rooflines, window placement, and signage. The curved mixed-use buildings facing the historic air station structure responds to its Modernism with extensive glazing and transparent canopies.

Pappageorge/Haymes Ltd.

600 North Lake Shore Drive
Chicago, Illinois

Above: Lake Shore Drive entrance.

Left: Paired towers flanking vine-clad garage wall.

Bottom left: Towers near left end in skyline view.

Occupying a pivotal position on Chicago's lakefront, this pair of towers is convenient to the urban activity of its Streeterville neighborhood yet offers the detachment of unobstructed lake views. On a site of about one acre, the project contains approximately one million square feet. Its 401 residential condominiums are divided into two slim shafts — of 40 and 48 stories — to limit the shadow cast on the neighboring beach. The concept of twin towers recalls Mies van der Rohe's world-famous paired apartment buildings nearby at 860-880 Lake Shore Drive, although the new pair is quite different in form and scale. Unlike Mies's rectangular towers, set back from adjoining streets, these building follow the boundaries of their site, reflecting the angular relationship of Lake Shore Drive to the prevailing Chicago grid. The tower walls facing outward consist of sleek expanses of blue-tinted glazing, divided into three-story horizontal bands. The walls facing inward to the 60-foot gap between the shafts present grids of punched windows in metal-paneled surfaces. The upper floors rise from a 500-car parking garage, the top of which is gardened between the towers, providing a shared sculpture garden and private terraces. The green theme is continued on the exposed walls of the garage between the towers, which will be clad with vines

Pappageorge/Haymes Ltd.

Block X
1145 Washington
Chicago, Illinois

Set in a context of industrial buildings and loft housing, this dense complex presents distinctly different images toward the streets outside and the court at the interior of its 1.77-acre site. Toward the streets, walls are regular, with large glazed bays echoing those of surrounding buildings. Toward the inner courtyard, volumes break down into a variety of balconies and projections, with exposed steel framing and fascias painted a lively blue. Structures containing a wide variety of flats and duplexes are pushed to the edges of the site to make room for the gardened court. The court – and the first floors of apartments – are a half level above grade, allowing for under-the-court parking, which is accessed from two streets without crossing pedestrian routes.

Top: Evening and daylight views of inner court.

Above: Street fronts echoing nearby buildings.

Left: Court fronts, with terraces stepping back between projecting stair towers.

Right: Court-level plan.

Photography:
Pappageorge/Haymes Ltd.
night view, The Thrush Companies.

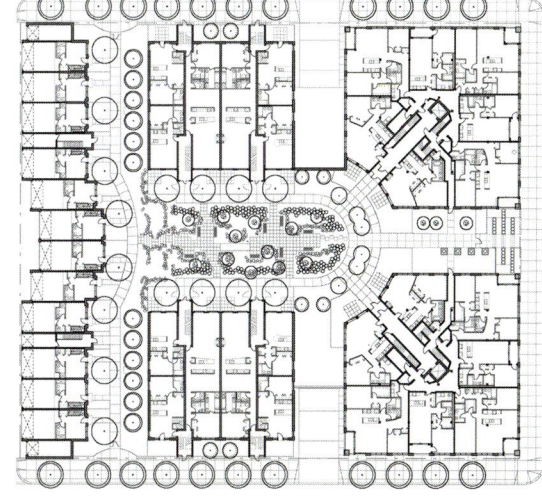

Pappageorge/Haymes Ltd.

Museum Park
Chicago, Illinois

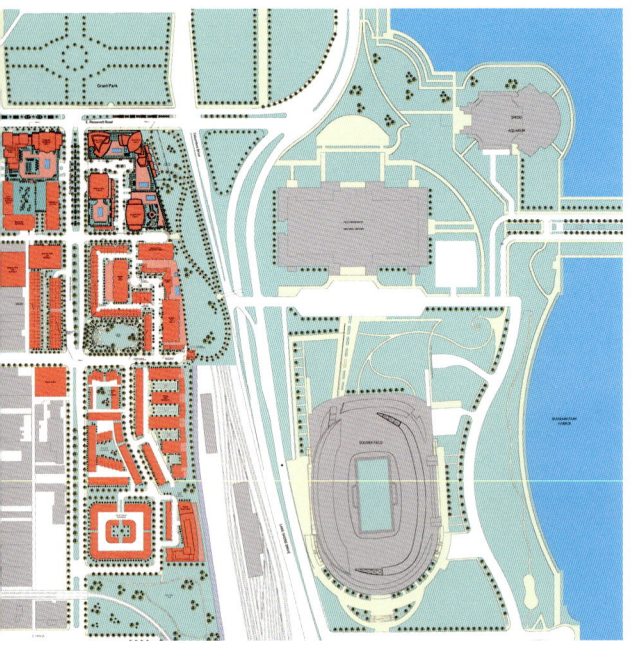

Until 1989, the 80-acre tract on Chicago's Near South Side known as Central Station was the vacant land of a former railroad yard. Adjacent both to Grant Park on the north and the recently reconceived Museum Campus on the east, the area presented a unique opportunity to create a new community responding to its prime location and the traditions of its surroundings. Occupying the largest part of Central Station, Museum Park includes a wide range of residential types: rental apartments, townhouses, condominiums, and loft units, along with a variety of community facilities, laid out along tree-lined streets and intimately scaled parks. To provide a strong sense of place and urban activity, all residential buildings front directly on streets. Generally speaking, taller buildings are sited toward the north and east edges of the site, facing large park areas and the lake, with low-rise buildings bordering established urban blocks to the west. Future construction proposed for Museum Park includes a 62-story residential tower with a distinctive organic prow-like form that is to rise at the northeast corner of the site, overlooking Grant Park, the Museum Campus, and Lake Michigan.

Top: Site plan.
Above: Townhouses facing a landscaped square.
Right: Residential loft structure.

Facing page, top left: Community recreational center.
Facing page, top right: Proposed residential tower at northeast corner of development.
Facing page, bottom: Large portion of project, seen beyond marina and Museum Park.

Photography:
Pappageorge/Haymes Ltd.

Pappageorge/Haymes Ltd.

Kinzie Park
Chicago, Illinois

Above: Townhouses facing Chicago River.

Left: Site plan.

Below: Mid-rise structure and townhouses at one end of site.

Below right: Passage to riverwalk between townhouse rows.

Photography: Pappageorge/Haymes Ltd.

The master plan and architectural design of this development take advantage of a prime waterfront property while dealing with its difficult edge conditions. The project's 4.3-acre site offers exceptional views of the Chicago River but is hemmed in by intensively used rail lines and major roadways. In response, the site plan organizes townhouses around two brick-paved internal streets, creating an enclave of traditional masonry and stone façades. Mid-rise apartment structures at the ends of the site buffer the development from train and traffic noises and offer broader vistas to their residents. Available to all residents is a riverwalk with carefully chosen plantings, benches, lighting, and sculpture. A central fountain plaza provides a pivotal link between the townhouse streets and the riverwalk.

Perkowitz + Ruth Architects

111 West Ocean Boulevard	Las Vegas, NV	Washington, DC
21st Floor	702.892.8500	703.668.0086
Long Beach, CA 90802		
562.628.8000	Portland, OR	Orange County, CA
562.628.8005 (Fax)	503.478.9900	714.850.3400
lglick@prarchitects.com	NW Arkansas	Studio One Eleven
www.prarchitects.com	479.271.8090	at Perkowitz + Ruth Architects
		562.901.1500
		www.studio-111.com

Perkowitz + Ruth Architects

Perkowitz + Ruth Architects

Bridgeport Village
Tualatin, Oregon

Above: Intimate passage evoking urban streets.

Below: Outdoor living room in central courtyard.

Right: View of main corridor lined with Italian kiosks.

Photography: Paul Turang Photography (above and right); Stephen Jones Photography (below).

Emerging from an abandoned rock quarry site, Bridgeport Village opened up the region to new high-end retail shopping opportunities. After a 20-year moratorium on large-scale retail development in the Greater Portland market, the new 500,000-square-foot open-air specialty village offers the latest in outdoor shopping concepts and technology to its regional consumers. Generating sales competitive with the top lifestyle centers in the nation, Bridgeport Village has become a place for friends and families to meet, shop and be entertained. The project emulates the urban nature of downtown with urban blocks, two story façades, and strategically placed narrow streets. As an extension of the urban architectural heritage of the greater metropolitan area, the design incorporates a wide spectrum of materials and façades from different eras. A fountain and play center inspire activity in the central courtyard, and an authentic imported gazebo is the site for outdoor afternoon concerts. Imported Italian kiosks line the main central aisle that ends with a large cinema, which serves as a dramatic yet elegant backdrop for the center. Fine dining options and coffee shops are easily accessible on outlying pads with curbside parking and valet options. Elegant landscaping and ambient lighting adorn the narrow meandering passageways. Complementary umbrellas and sheltered areas provide rain protection. Strategic placement of a parking structure behind the village places 1,140 cars in close proximity to shops.

Studio 111
at Perkowitz + Ruth Architects

Infill & Mixed Use

Studio One Eleven is an architecture and urban design practice dedicated to smart growth principles as an alternative to conventional suburban development models. From high-rise housing, mixed-use infill, hospitality and civic buildings to downtown revitalization, streetscape and storefront design, each endeavor aims at making a walkable, humane and sustainable larger urban whole.

Figueroa Central Downtown Los Angeles, California

Located adjacent to the Los Angeles Convention Center, Staples Center and the future L.A. Live and Nokia Theater developments, Figueroa Central, designed in collaboration with Johnson Fain International Inc., brings an ambitious program of retail and residential uses into the city's burgeoning downtown. With two residential towers defining the skyline, the development focuses flagship and neighborhood retail with restaurant uses along sidewalks, plazas and paseos to engage the public realm and avoids the street-deadening effect of inward-facing retail promenades. An environmental graphics program contributes to an exciting atmosphere reminiscent of Times Square.

Right: Aerial view highlighting the project's integration within the emerging entertainment district.

Studio 111
at Perkowitz + Ruth Architects

Above: Central green of Echelon is defined by a formal building edge.

Above right: Passive cooling tower, which captures breezes and cools them as they descend into intimate courtyards of Echelon, allowing for outdoor activities year round.

Echelon I, Las Vegas, Nevada

Echelon I is a 15-acre, 372-unit component of a larger residential campus. Buildings are arranged around linear quads and courts to maximize outdoor living and views of the Las Vegas strip and surrounding mountains. In response to the hot dry climate, the orientation of the buildings, cooling towers, trees and fountains are designed to cool and shade courts. To mitigate the desert heat gain, adjustable canopies, louvers and screens allow residents the ability to control individual environments.

Milan Lofts, Pasadena, California

Located within walking distance from Old Town Pasadena and Del Mar Station, Milan Lofts is designed as four buildings arranged around communal courts. Through massing and stylistic variety the mixed use transit oriented development integrates into the fine grain of the surrounding urban fabric.

Above: Development of Milan Lofts is consciously articulated as several buildings.

Perkowitz + Ruth Architects Mercantile West
Ladera Ranch, California

Left: Gateway to development's Main Street, which can be closed to traffic for community events.

Below: Shop fronts evoking individual buildings on a traditional commercial street.

Photography: Paul Turang Photography.

This 320,000-square-foot development is the community shopping center for Ladera Ranch, a 4,000-acre planned community in Orange County. The 14.5-acre project combines functions of the neighborhood center and a pedestrian-oriented "Main Street." Retail shops want the intimacy of a Main Street, while restaurants, pharmacies, and supermarkets prefer outparcels with the best accessibility by car. In this case, conflicting needs were reconciled by laminating the supermarket with Main Street storefronts. The Main Street itself, flanked by shops, with diagonal parking along a central linear island, is located for convenient pedestrian and bicycle access from an adjacent community park and residential area. Storefronts in a variety of natural materials recall the architecture of the mid-century downtown core of traditional California communities.

Perkowitz + Ruth Architects

Buena Park Downtown
Buena Park, California

Above: Site plan.

Right and below: Central plaza under different lighting conditions.

Photography: Paul Turang Photography.

Once a successful, though typical, suburban mall, the development saw its market slip away as other destinations attracted its customers. It has now been reinvented as a vibrant town center with 138,000 square feet of new or renovated building space, adding a strong entertainment component to the revitalized shopping facilities. The design encourages pedestrian flow between the existing enclosed mall and the open-air zone, with the prominent theater lobby sited on axis with the main mall entry. Perkowitz + Ruth Architects created an urban setting that blends Modernist and Neo-Traditional styles to infuse the feeling of a place developed over time. Many amenities in the public open spaces enhance the community feeling. The central plaza combines a series of elegant fountains, a reflecting pool, seating areas, a café kiosk, and large shade trees.

Perkowitz + Ruth Architects The Lakes at Thousand Oaks
Thousand Oaks, California

Above: Project elevation that highlights strong tower element, alternating facades, and dramatic landscape and lighting.

Left: Perspective of the pedestrian experience.

Photography: Paul Turang Photography.

The Lakes is a newly opened lifestyle destination created by the collaborative talents of Caruso Affiliated, Perkowitz + Ruth Architects, Lifescapes International Inc. and Francis Krahe Lighting Design. This refreshing outdoor location offers multiple options for family entertainment, featuring a mix of restaurants and specialty stores. The Lakes also has a seasonal ice rink incorporated into a new lake replete with playful water features.

The building design offers alternating façades orchestrated along the perimeter of the lake. Embedded within a sophisticated interplay of elegant landscape design and pedestrian paseos, the integration of dining opportunities and specialty shops will draw patrons to this new, stunning destination.

Retzsch Lanao Caycedo Architects

137 West Royal Palm Road
Boca Raton, FL 33432
561.393.6555
561.395.0007 (Fax)
www.rlcarchitects.com

Retzsch Lanao Caycedo Architects

Royal Palm Office Building
Boca Raton, Florida

Top: Building seen from street.
Above: Lobby.
Right: Identifying tower.
Facing page: Trellis.
Photographs: Chuck Wilkins Photography.

The distinctive architectural tradition of Boca Raton, established by Addison Mizner and his contemporaries in the 1920s, has been respectfully reinterpreted in this 5,600-square-foot office building. Located in the city's redevelopment district, the building contributes to its urban character by defining the street edge and recapturing the porch, a typical Florida architectural feature. The towers at the building's corners enhance its one-story scale and lend it a consistent identity when seen from its most visible sides. The wood trellis filters daylight entering the interior and casts an intricate pattern of shadows on exterior walls. A plaza at the rear of the building, framed by the building envelope and lush landscaping, serves as a gathering place and entry court not unlike the traditional courtyards of Mediterranean architecture.

Retzsch Lanao Caycedo Architects

Cypress Park West, Phase II
Fort Lauderdale, Florida

Above: New structures reflected in pond.
Below: Lobby interior.
Bottom: Walkway canopy.
Bottom right: Entrance façade.

Facing page: Aerial showing garage, office building, and new landscape design.
Photography: Chuck Wilkins Photography.

The second phase of Cypress Park West has transformed a single office building into a true office park, attracting a major tenant that might otherwise have moved its large workforce out of the Fort Lauderdale area. A new 116,000-square-foot office building and the existing building, renovated to match its class A standard, now accommodate Microsoft's Latin American headquarters and Nextel offices. The redesigned five-acre site now pulls together the two office buildings, a full-service Marriott, and a six-level parking garage into an attractive and functionally effective composition. New construction is sited well away from an adjacent eight-acre nature preserve to minimize impact on its natural growth and wildlife. The renovation of the Phase I office building was accompanied by the redesign of its plaza, the latter enhanced as an entrance to the development by royal palms relocated from other parts of the property. Lighting was designed to create a campus atmosphere, with bollards along walkways and plant lighting in the plaza. Construction was completed on time and within budget. Tilt-up concrete walls for the office buildings yield smooth white surfaces that contrast appealingly with green glass and surrounding foliage. The garage was constructed with twin tees and other precast components.

Retzsch Lanao Caycedo Architects

Fifth Avenue Place, Phase II
Boca Raton, Florida

Building on the success of the firm's Fifth Avenue Place development, completed in 2001, the proposed second phase will add the nine-story Fifth Avenue Tower. The 52,995-square-foot structure will have retail and restaurant space on the street floor, a recreation deck with a community pool on the second-floor setback, and condominium units above. An urban infill project, Fifth Avenue Place is located in the community redevelopment area of downtown Boca Raton, between a commercial area and a single-family residential district.

The first phase included townhouses that maintain the street line along Boca Raton Road and serve as transitional elements between low- and medium-rise neighbors. A parking structure in that phase of construction was located behind the townhouses and partially topped by two floors of offices, which offer views of the ocean and the Intracoastal Waterway. The new tower is expected to be completed in 2006.

Left: Phase I complex.

Below: Plan and elevation of phase I and II.

Facing page: Rendering of proposed tower.

Photography: Chuck Wilkins, phase I.

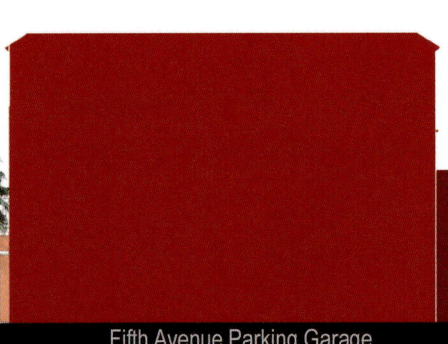

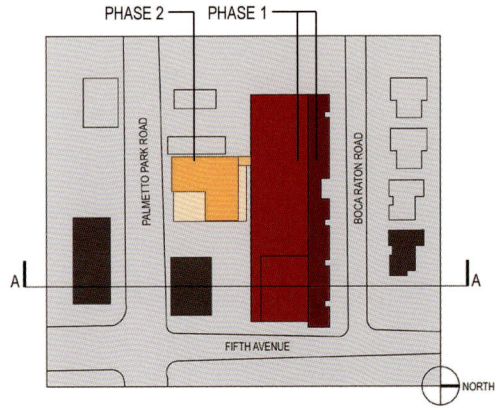

Retzsch Lanao Caycedo Architects

The Pointe at Middle River
Oakland Park, Florida

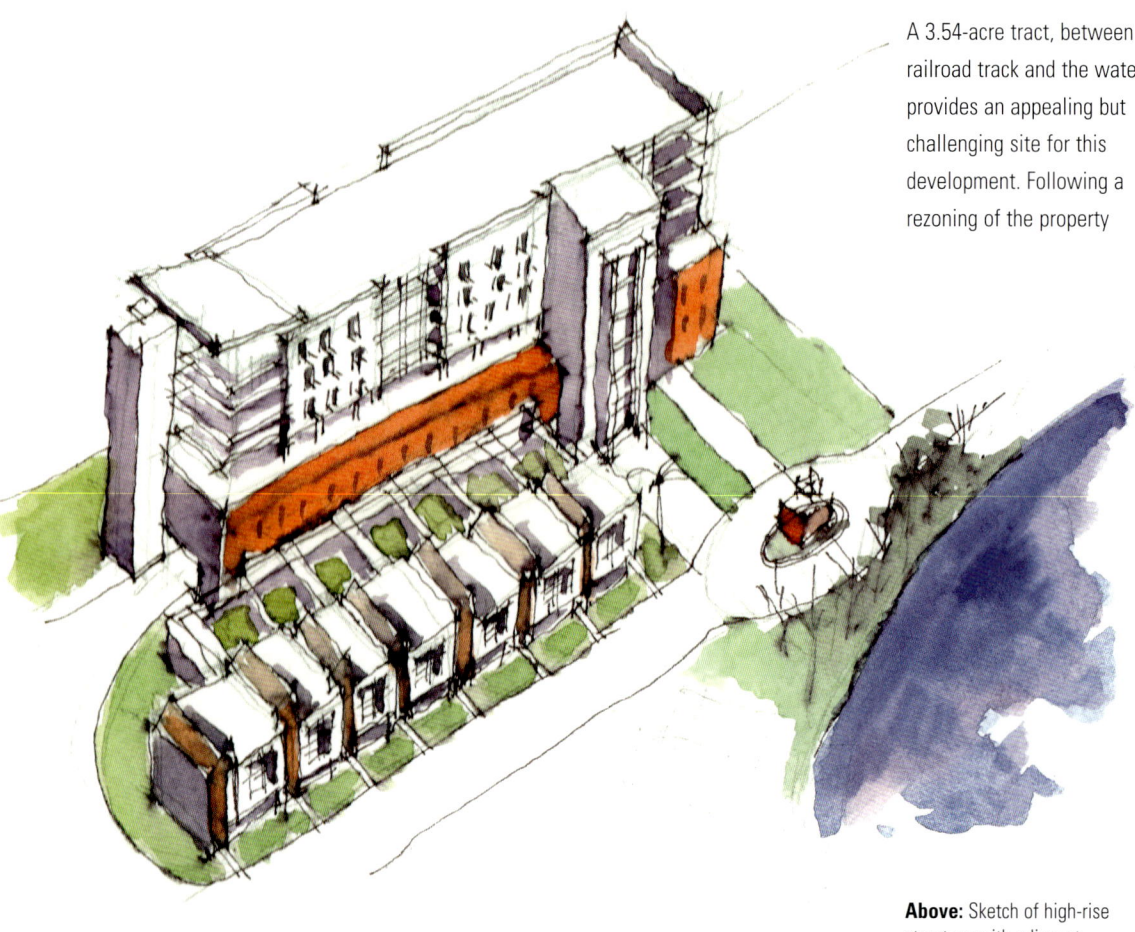

A 3.54-acre tract, between a railroad track and the water, provides an appealing but challenging site for this development. Following a rezoning of the property from business to residential, the architects organized the development with a high-rise residential structure rising above a three-story parking garage along the railroad side, shielding the rest of the development from rail activities. The 68 one- and two-story loft units in this building will offer views out to the water above the 37 three-story townhouses, with two-car garages, that extend out to the water's edge. The open site area of 92,322 square feet will include a clubhouse and pool. Construction will be of post-tensioned concrete, with CMU walls, some of which will be highlighted in vivid colors. The project is scheduled for completion in 2008.

Above: Sketch of high-rise structure with adjacent townhouse units.

Below left: Site plan.

Below right: Elevation of high-rise.

RTKL

Baltimore 410.537.6000	Chicago 312.704.9900	Shanghai 86.21.6122.5922
Dallas 214.871.8877	Miami 786.268.3200	Madrid 34.91.426.0980
Washington 202.833.4400	London 44.207.306.0404	
Los Angeles 213.627.7373	Tokyo 81.33583.3401	www.rtkl.com

RTKL

LaQua Tokyo Dome City
Tokyo, Japan

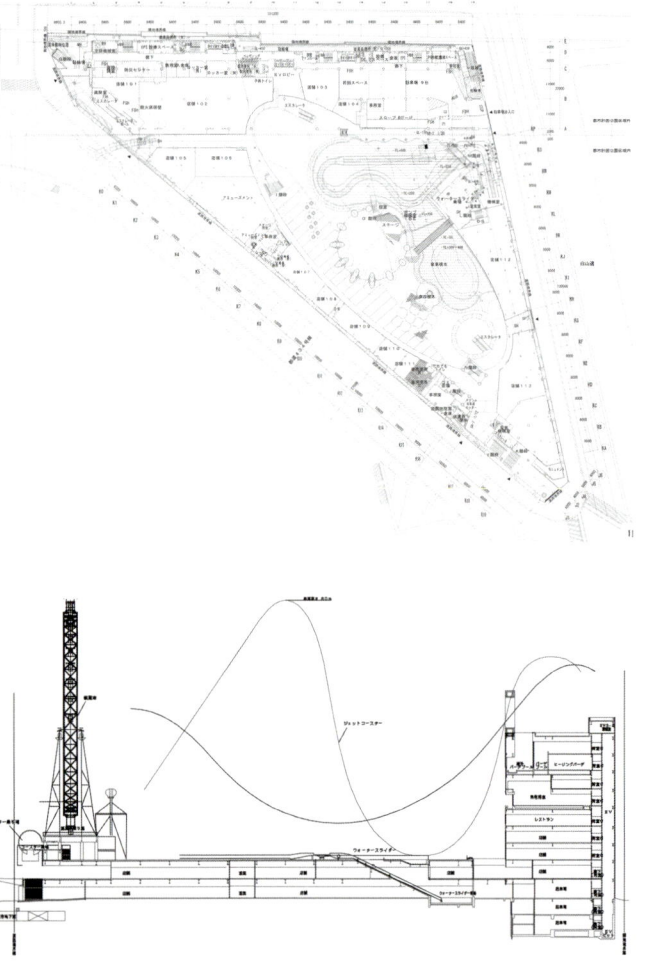

Fun is visibly the function of this entertainment center adjoining a domed arena in Tokyo. An iconic ferris wheel and a roller coaster that snakes through the complex are the most visually compelling elements of a development that includes a water slide, a haunted house, a concert stage, a spa, and over 70 shops and restaurants. Zoning requirements stipulated that the main structure be located at the back of the site, along a railroad line. There a multi-story building houses retail and dining facilities on the lower levels and a state-of-the-art spa on the three upper floors. The architects' objective was to weave the active and passive functions together in a single coordinated experience. The outdoor rides and their support spaces are laid out around a roughly elliptical landscaped court. Subtle cues of lighting and design create segues from intensely active areas such as the roller coaster to the tranquility of the spa. A design challenge was integrating the necessary, but traditionally unsightly, safety accommodations of the rides into the project's aesthetic. The solution included designing the catchments for the ferris wheel to look like ships' sails and crafting safety barriers from stained wood to recall the hulls of historic ships.

Above left: Plan and section of complex.

Left: Central court and nautical-looking slab building.

Right: Entertainment center with domed arena at left.

Bottom right: After-dark view.

Photography: Courtesy of Takenaka Komuten KK.

RTKL

Downtown Brea Development District
Brea, California

Approached to develop a master plan for an ambitious mixed-use project in the Southern California city of Brea, RTKL collaborated with municipal planning and development agencies and the city council to design a district with memorable identity and market flexibility. The project occupies 25 acres in the heart of Brea's downtown and includes 220,000 square feet of retail and restaurants, nearly 100 residential units of various types, parking structures, and civic spaces, supplemented by a year-round events program. Its neo-traditional master plan recalls the ambiance and round-the-clock vitality of a classic Main Street. Pedestrian amenities include street trees, seating areas, fountains, and public art. On-street parking protects pedestrians and slows traffic. A district-wide parking authority allows shared parking solutions, thus reducing total demand requirements. Design criteria for building design encouraged individual expression for commercial properties, while establishing a distinctive overall character for the district and integrating key civic elements that celebrate Brea's unique history.

Left: Gateway to the district.

Below left: Aerial view of street.

Below: Night view with lively signage.

Left: Sidewalk activity and signs in pop tradition.
Right: Pedestrian environment.
Photography: RTKL Associates.

RTKL

Principe Pio
Madrid, Spain

A new 110,000-square-foot retail development has been created at an existing transportation hub in the dense core of Madrid. The project juxtaposes 70 high-end shops, a multi-screen cinema, and restaurants to a complex of rail, subway, and bus facilities, plus an 840-car underground garage. All transportation lines had to be kept running during construction. Architecturally, the clearly modern elements had to be integrated with the existing train station to form a cohesive environment. By moving several train tracks out of the existing station, it was possible to enclose the area under the old canopy that had sheltered the platforms. Innovative engineering techniques allowed excavation below this canopy to make space for a multilevel retail center with parking under it. A new central glass dome, with a different but related aesthetic, serves as an entranceway and a connection between the reused railroad station spaces and a new entertainment wing anchored by a cinema complex. Principe Pio is one of the most successful retail projects in Spain, with an average of 60,000 visitors monthly, and is 100 percent leased at rates 20 percent above the Madrid average.

Above: New entertainment wing and glass-dome entry.
Below: New parapet under old station canopy.
Below right: Multilevel retail inside historic shell.
Facing page: Domed rotunda, retail beyond.
Photography: David Whitcomb, RTKL.

RTKL

Zha Bei/The Hub International Lifestyle Centre
Shanghai, China

For an underdeveloped area of Shanghai still showing the scars of World War II bombings, this project will provide an essential mixed-use urban center. The 55,422-square-meter (599,000-square-foot) site will accommodate nine buildings totaling 202,199 square meters (2,184,000 square feet), with 8,000 square meters (86,400 square feet) of public open space. The mix of functions will comprise offices, an entertainment center including a cinema, a hotel, residential and mixed-use lofts, shops, and restaurants, all linked together by a network of pedestrian-friendly streets and extensively landscaped open spaces. Care has been taken to integrate the new development with the existing urban fabric so that it becomes a part of everyday local life, rather than an enclosed and alien presence.

Above: Aerial view of complex.
Left: Evening view along street.
Below left: After-dark aerial.
Below: Shops around landscaped court.

Sasaki Associates, Inc.

64 Pleasant Street
Watertown, MA 02472
617.926.3300
617.924.2748 (Fax)
info@sasaki.com
www.sasaki.com

77 Geary Street
Fourth Floor
San Francisco, CA 94108
415.776.7272
415.202.8970 (Fax)
sanfrancisco@sasaki.com

Sasaki Associates, Inc.

Addison Circle Park
Addison, Texas

Facing page, left: Transverse colonnade that bounds main performance area.

Facing page, right: Performance in progress.

Below: Fountain pool with existing sculpture by Mel Chin in roundabout beyond.

Top: Aerial view of park during a major event, showing high-density housing along some bounding streets.

Above: Central portion of park, showing transverse colonnade.

Left: Fountain near residential area.

Photography: Paul Chaplo (aerials), Craig Kuhner (others).

Located near the Dallas North Tollway, the park is the focus of a rapidly growing high-density community. For this key open space, the town wanted an inspiring urban park that would be ideal for the everyday use of residents yet be adaptable as a setting for regional festivals of over 100,000 people. Initial planning set the boundaries of its 10-acre site and located a new street with parcels for future mixed-use development along a light-rail transit corridor to the south. Numerous meetings with the town's events coordinator and potential operators determined the park's technical needs. The practicalities of public events were considered in the layout and regrading of the flat land for good sight lines, while preserving key trees. Provisions were made for a raised stage — allowing for operators to set up appropriate stage configurations and lighting for specific events. Locations of tents for food and beverage vendors were established, and underground electrical, water, and sewer hook-ups and paved ramps were provided. A pavilion building, with kitchen, restrooms, and sheltered seating is designed as the headquarters for events and can be rented out for wedding receptions and other festivities. The resulting park can accommodate such celebrations as an annual Octoberfest as well as intimate jazz and classical concerts.

259

Sasaki Associates, Inc.

Thu Thiem New Urban Center
Ho Chi Minh City, Vietnam

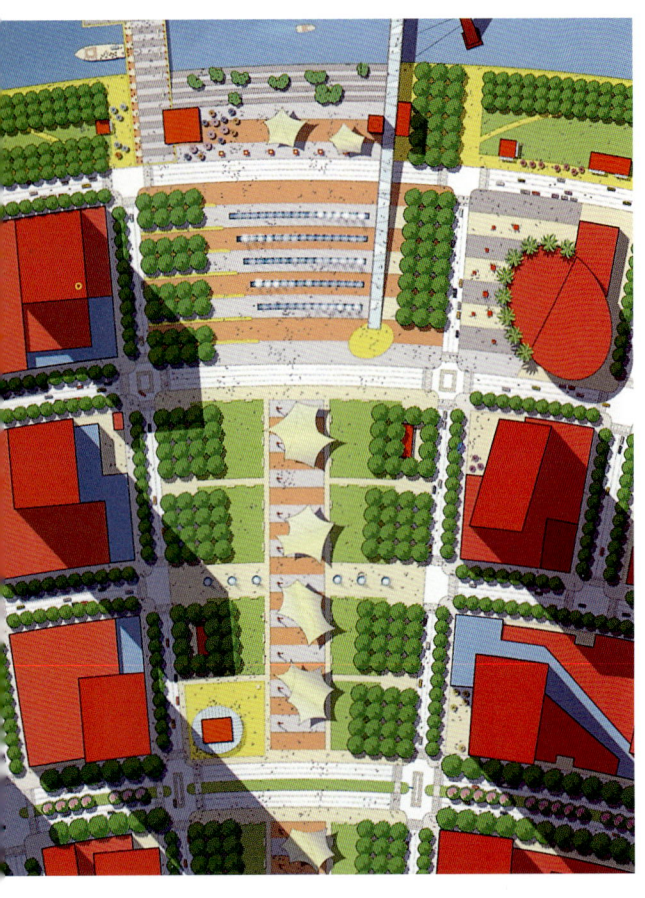

With an ultimate build-out of 68 million square feet on 1,840 acres, this new urban district is sited across the Saigon River from the center of the existing city. It will ease development pressure on the historic core and transform the river — now largely an industrial "back door" — into an important asset. The new development will feature waterfront open spaces and promenades. An iconic pedestrian bridge will connect a new central plaza to Me Linh Square in the old city. The new master plan responds to the bend in the river, and its primary road, Crescent Boulevard, echoes the curve. The plan recommends that a proposed highway through the district become an urban boulevard, with major traffic diverted onto a regional expressway. A system of canals and lagoons running through the new area will connect directly to the river. Areas of flood plain will be retained for tidal and storm-water management. The high density of built areas will favor pedestrian and transit access.

Approximately 60 percent of the new district's housing and other facilities will be within a 15-minute walk of the central plaza.

Facing page left: Detail plan of new central plaza, with end of footbridge and flanking buildings.

Facing page right: Thu Thiem hydrology concept.

Facing page below: New riverfront, with end of vehicular bridge in foreground; footbridge linking plazas in new and existing districts in middle distance.

Right, top: Urban design plan of the mixed-use core area district.

Right, bottom: Model of New Urban District.

Sasaki Associates, Inc.

Charleston Waterfront Park
Charleston, South Carolina

Charleston is a city of rare charm with an exceptional architectural heritage, yet the decline of its older port facilities had left vacant properties and rotting piers within a ten-minute walk of its downtown core. To revitalize the historic area, Charleston has carried out a long-term plan to reclaim the waterfront for the public. The Waterfront Park at the foot of Broad Street is part of the mayor's vision of a continuous promenade along two rivers, and is meant to redistribute some tourist activity from other heavily traveled parts of the core. Engaging key street corridors, the park allows views of the water and is designed as a destination place of civic stature. Its main entrance leads to an interactive fountain on a plaza that serves as a forecourt to a new wharf extending 365 feet into the Cooper River. A raised lawn scaled for large gatherings centers on another fountain, which takes the form of a pineapple, a traditional symbol of Southern hospitality. Bosques of live oaks along the city edge provide welcome shade. Restored salt marshes along the water's edge support wildlife and give visitors an appreciation of the area's marine ecology. Federal regulations required that publicly accessible parts of the park be raised 6.5 feet to the 100-year flood level. Unstable site conditions necessitated state-of-the-art stabilization of the landfill and 60-foot piles to support the pineapple fountain.

Left: Overall view of park.

Above: Entry plaza fountain in foreground, live oak grove, lawn centered on pineapple fountain, and palmetto-lined promenade.

Above right: Pineapple fountain in action.

Right: Portion of riverfront promenade.

Below right: Adger's Wharf gardens.

Photography: Landslides (left), David S. Soliday (above), Sasaki Associates (others).

Sasaki Associates, Inc.

Detroit Riverfront Civic Center Promenade
Detroit, Michigan

The Promenade meshes an active waterfront with recreational uses to link the city and its riverfront. Extending 3,000 feet along the Detroit River and a mere 55 feet wide in places, the site had previously been used mainly for surface parking. Key elements of the plan are: a grand stair linking Hart Plaza, the city's major public space, with the promenade; historic plaques marking Cadillac Landing, where the city was founded; and parallel bands of lawn recalling the area's 18th-century "ribbon farms." The 12-foot-high helix platform, reminiscent of a mariner's coiled rope, unwinds to form the serpentine seatwall. Undulations in the seatwall embrace the 20-foot-tall columns of a public transit viaduct, helping to integrate it with the promenade. Concave segments of the seatwall accommodate seating, and sections of the steel railing can be removed to accept gangplanks from visiting vessels.

Above: Downtown Detroit with promenade in foreground, bordering Joe Louis Arena and COBO Conference and Exhibition Center.

Photography: Landslides (aerials), Christopher Lark (others).

Below left: Corner of promenade showing night lighting and visiting vessels.

Below center: Close-up of seating helix and undulating seatwall.

Below right: Boats tied up along modular railing.

SEH (Short Elliott Hendrickson Inc.)

100 North 6th Street
Butler Square Building
Suite 710C
Minneapolis, MN 55403
612.758.6700
612.758.6701 (Fax)
866.830.3388 (Toll free)
www.sehinc.com

SEH (Short Elliott Hendrickson Inc.)

Mound Public Safety Facility
Mound, Minnesota

Left: Main entrance facing intersection.

Below: Vehicle bays flanking identifying tower.

Photography: ©Bob Perzel.

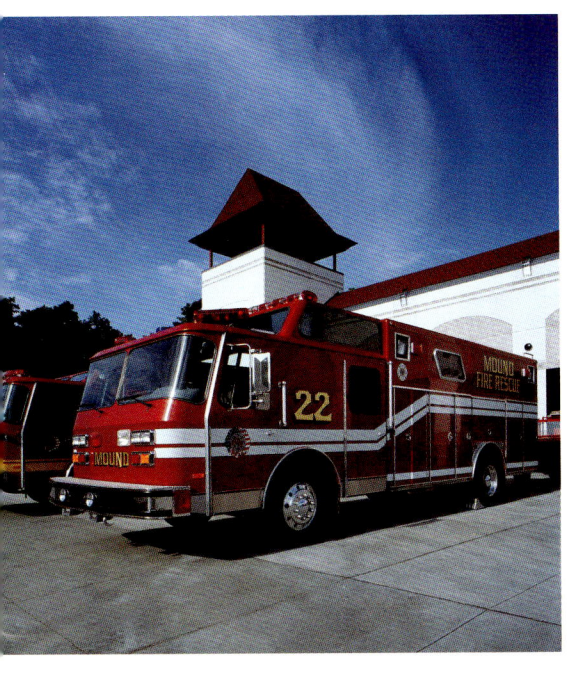

Left: Fire equipment on front apron.

Right: Fire station interior with exhibit wall and vintage fire truck.

Below right: Details of precast concrete exterior panels and balcony railing.

On a 3.72-acre site, the architects have combined fire and police stations in a way that offers the economy of shared facilities, yet maintains a clear-cut identity for each. A subtly night-lit tower announces the structure from a distance. A bold curved front, facing a major intersection and approached by a flagpole-adorned podium, indicates the main public entrance. Inside, the fire station occupies 23,084 square feet, the police station, 12,000 square feet. For public events, a commercial kitchen can provide meals for 5,000 people over a four-hour period. A glazed arcade on the front of the building provides a base for an open balcony facing the street. Exterior panels of precast concrete give the structure an appropriately dignified yet unpretentious image.

SEH (Short Elliott Hendrickson Inc.)

Heart of Anoka Commuter Rail Village
Master Plan
Anoka, Minnesota

The objective for this 150-acre site was development that will support future commuter rail but is not dependent on it for its success. The resulting plan is based on solid market research and deep understanding of community preferences regarding development and design. Railroad station program elements such as parking and support services are integrated into mixed-use commercial and residential buildings. Medium- and high-density housing is organized around the station, making compatible transitions at the site boundaries to existing residential areas. Some existing buildings would be retained, including one industrial structure converted to loft housing units. Streets and bridges would be enhanced, new grade-separated railroad crossings created, and the riverside trail system extended. The community will demonstrate the possibilities that exist where there might otherwise be just a railroad platform and a parking lot.

Facing page, top: Aerial rendering of development.

Facing page, bottom: Residential blocks.

Above: Master plan, highlighting specific uses.

Right: Commuter station and related mixed-use buildings.

SEH (Short Elliott Hendrickson Inc.)

Loring Bikeway and Park
Minneapolis, Minnesota

The bikeway was needed to make a cohesive connection for cyclists and pedestrians across complex street and highway geometries, using public right-of-way. The bridge and the remainder of the one-mile-long bikeway have been designed to respond to their context, including the cultural and architectural traditions that define the neighborhood. The bridge will follow a long curve in plan, with visually engaging barriers on either side. Joining it at either end will be linear trails threaded through and aligned with the predominant street grid. A pocket park is planned for a vacant, neglected triangular lot adjoining the bikeway. Its green space will provide a rest area for trail users and a gathering spot for local residents, and a perennial garden within it will serve as a neighborhood activity. The project had to meet various standards of the state and federal transportation departments as well as accessibility requirements. The city, county, and state contributed to its $1.5 million cost.

Above: Bridge with decorative screen wall and railing design.

Below left: Bicyclists on bridge access ramp.

Below: Plan of pocket park serving also as trail head for bikeway.

Below right: Photographic map of area with trail and bridge routes superimposed.

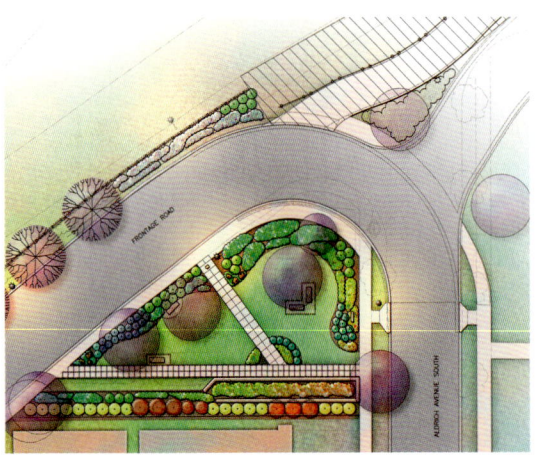

SEH (Short Elliott Hendrickson Inc.)

I-35W Access Project
Hennepin County, Minnesota

This freeway improvement project will help mend the urban fabric so drastically damaged by the highway's original construction. When the I-35 corridor was slashed through 70 blocks in the 1960s, it divided neighborhoods, obstructed access, and contributed to the concentration of poverty in some areas. Led by a partnership of landscape architects, the I-35 Access Project grew out of years of input from neighborhood residents and business and institutional groups. The focus is on mitigation of the freeway's negative effects and enhancements to the public realm, including noise abatement and integration of public transit. Adjacent redevelopment parcels will be developed with neighborhood-compatible structures. A comprehensive streetscape and traffic-calming program is included for adjacent city streets. The highway bridges will be opened to local streets below, letting in light and air and providing opportunities for public art and fountains. Along a one-mile stretch, the center lane will be used for bus rapid transit, with multi-level stations to bring passengers from the city streets to the freeway level. The design had to meet federal and state highway and bridge standards and federal guidelines for context-sensitive design. The project does more than put a new face on the mistakes of the past. It builds a strong case for public involvement in creating wider freeway crossings, with inviting pedestrian lanes, as well as new landscaping and parks along and adjoining freeway routes.

Above: Elevated highway with masonry towers for transit station access.

Right: Proposed highway overpass and bus rapid transit station.

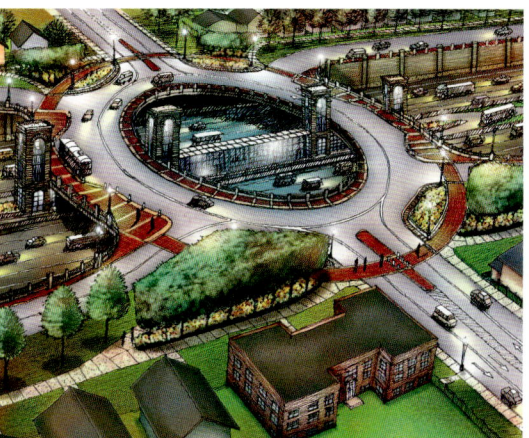

Far left: Early alternative for bridge over I-35 at 38th Street, showing pedestrian walkways, landscaping, and public art.

Left: Final, preferred "Ellipseabout" design for 38th Street bridge, with bus rapid transit station below.

SEH (Short Elliott Hendrickson Inc.)

Gateway Centre
Longmont, Colorado

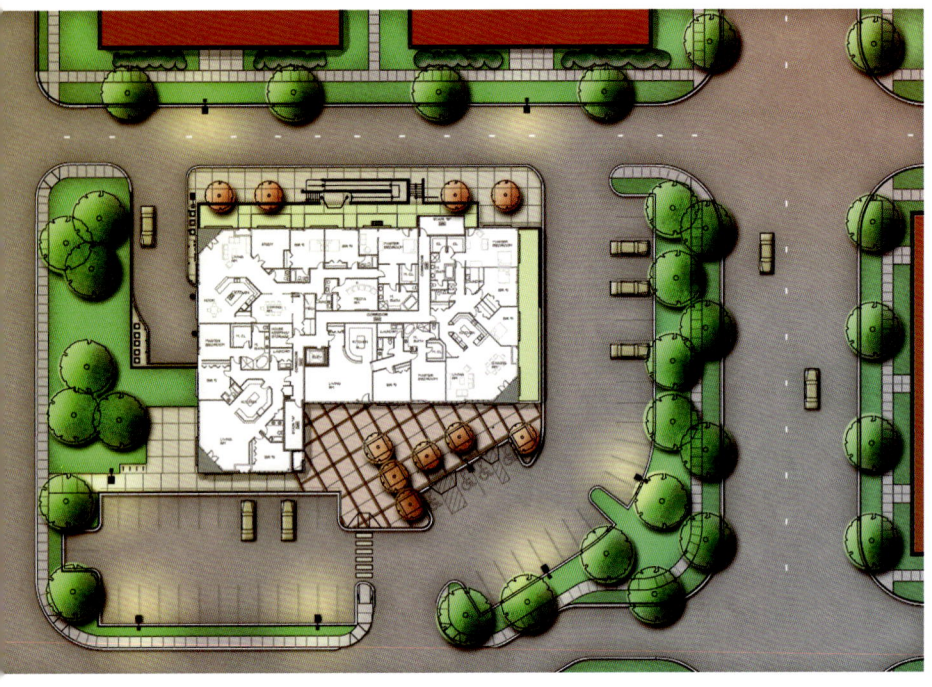

Above: Plan of residential third floor.

Below: Four elevations, with main street front at lower right.

This development will introduce residential units as one element of a mixed-use program on a site where its immediate neighbors are all commercial. The project will contain retail and offices on the street floor, offices on the second, and luxury lofts on the third. Exterior walls will be a combination of concrete block, brick, and stucco. The building entrance, reached by way of a paved and planted front terrace, will be identified by a two-story-high glazed wall. Three of these lofts will have corner balconies, defined by the 45-degree geometry of their living rooms. Residents will enjoy views of the Front Range of the Rockies, visible from upper floors at this location. Loft residents will have basement storage areas and private parking spaces in the garage under the building. Spaces there will also be available to some commercial occupants.

SWA Group

2200 Bridgeway Boulevard
PO Box 5904
Sausalito, CA 94966
415.332.5100
415.332.0719 (Fax)
www.swagroup.com

Sausalito
Laguna Beach
Houston
Dallas
San Francisco
Los Angeles
Shanghai

SWA Group

Lite-On Electronic Headquarters
Taipei, Taiwan

For its corporate headquarters, Lite-On wanted its building and landscape integrated to create a "green" complex. As soon as the architects were commissioned, they brought in SWA to work closely with them. The concept developed by this team was to place the private work spaces in a 25-story tower and the public facilities in a podium with a gardened roof. Under the podium is a four-level below-grade garage. The gardened plane became the focus of the project, sloping from the second level down to street level toward Gee Long River. Garden planting had to be chosen to flourish in shallow soil, areas of deep shade, and exposure to winds. Watercourses are organized to collect storm water, which is stored and used for irrigation. The direction of view from the building toward the river is emphasized with linear planting beds and walkways that bridge over a sunken light well/courtyard planted with camphor trees.

The tower-and-podium concept gave the project a distinctive image in its crowded urban area, opening up welcome space and views. The owner's endorsement of environmental principles yielded the first green roof built by a private developer in Taipei.

Above: Planted rooftops.

Left: On-site waterfall.

Right, top to bottom: Sign at road; building from road; entrance sunken into podium; pools cascading toward river.

Facing page, top: Aerial view of landscaped podium roof.

Facing page, bottom: Bridges where sloping landscape crosses sunken courtyard.

Photography: Tom Fox.

275

SWA Group

PPG Place
Pittsburgh, Pennsylvania

Left: Computer-controlled jets can vary from 2 in. to 15 ft.

Near right: New honey locust with original building details.

Middle right: Plaza at night, with programmed fountain and lighting; water jets as photo backdrop.

Far right: Plaza in normal summer mode.

A failing plaza at a landmark city-center complex has been transformed into a success. Recognized as an icon of Post-Modernism from its opening in 1984, PPG Place is a 5.5-acre, 1,570,000-square-foot development composed of six Neo-Gothic glazed towers topped by 231 glass spires. While its skyline image is powerful, its central plaza, a half-acre of paving with no trees, constructed over a parking garage, turned out to be forbidding and unpopular. In 2001 PPG Place's new owners commissioned SWA to reinvent the plaza as a pedestrian destination that would attract upscale tenants to the complex's street-level retail arcade and encourage investment in surrounding blocks. Key to making the plaza attractive are locust trees in stainless steel planters, outdoor dining areas under 12-foot-wide umbrellas, and a 140-jet computer-choreographed fountain. Since the tree planters are movable and the water jets flush with the pavement, the plaza can be quickly cleared for the numerous concerts and festive events scheduled there. In the winter, a 9,586-square-foot skating rink is installed around the 40-foot central obelisk, using low-tech layers of sand and Styrofoam to level it on the sloping surface. Philip Johnson, renowned architect of the PPG Place buildings, judged the revitalized plaza "marvelous."

Left: Dining tables and movable tree planters.

Right: New water jets around orginal 40-ft. obelisk.

Photography: Tom Fox.

SWA Group Hangzhou HuBin Commerce & Tourism District Redevelopment Master Plan
Hangzhou, China

A city of 3.3 million people, Hangzhou is located on the shore of West Lake, an internationally recognized recreation and tourist destination with ancient temples, teahouses, and moon bridges. But a major thoroughfare separated the lake from the core of the city. The first phase of a master plan, completed in 2003, has transformed the relationship. One key strategy was construction of a 1.5-kilometer (0.93-mile) tunnel beneath the lake, diverting traffic from the lakefront road, which has been redesigned as a pedestrian-friendly boulevard bordering an expanded waterfront park. "City Stream," a pattern of water courses, still pools, and waterfalls passing through mid-blocks, courtyards, and plazas, recalls rivers that once flowed through the district. Historic buildings have been preserved and adapted as cultural and art centers, and a new 10,000-square-meter (108,000-square-foot) focal plaza overlooking the lake has been created.

Top: Lakefront boulevard and park.

Above, left and right: "City Stream" waterway; boulevard in use.

Far left and left: Commercial district with new lighting.

Photography: Tom Fox.

SWA Group   Lewis Avenue Corridor
Las Vegas, Nevada

Above: View along avenue.
Below: Two views of stream bed.
Bottom: Quotations from local poets and writers at footbridge.
Right: Rock-rimmed pool, federal courthouse beyond.
Bottom right: Waterfall "source" at courthouse.
Photography: Tom Fox.

In a city known for fantasy, the design of Lewis Avenue affirms the real-world nature of its district. As part of a downtown revitalization effort, the city asked SWA — already engaged in two major projects along the avenue — to redesign a three-block corridor linking key governmental buildings. At the outset, the avenue presented a four-lane roadway and narrow, unshaded sidewalks. Working with city departments, SWA found that two traffic lanes could be eliminated, making room for 20-foot sidewalks with double rows of trees. It was possible to remove surface parking from one key block and turn it into a plaza.

Influenced by the rugged regional landscape, the redesign places trees much as they would appear at an oasis and uses water sparingly as in a desert wash. A dramatic 10-foot waterfall from the level of the federal courthouse entrance is the apparent source of the stream below.

SWA Group

Santana Row
San Jose, California

For the development of a neo-traditional town center near downtown San Jose, SWA provided landscape architectural services. At full build-out, the 42-acre site will contain 680,000 square feet of retail space, including shops and boutiques lining the main street, 1,200 residential units, a 231-room luxury hotel, 15 to 20 restaurants, and open-air cafes, along with parks, plazas, and landscaped streetscapes. Apartments, condominium units, lofts, and the hotel are located above street-level commercial spaces. The design intention, for both buildings and landscape, was the use of a variety of styles to create the impression of growth over time. The project included complete upgrading of infrastructure, transportation circuits, parking, access to nearby highways, and wireless internet access throughout.

Top: Ample outdoor dining areas.

Above: Planting as traffic calming device.

Left: Active plaza with passage into courtyard beyond.

Below: Typical streetscape and signage.

Far left: Streetscape with intended "European" character.

Photography: Tom Fox, Bill Tatham.

Swaback Partners pllc

7550 East McDonald Drive
Suite A
Scottsdale, AZ 85250
480.367.2100
480.367.2101 Fax)
info@swabackpartners.com
www.swabackpartners.com

Swaback Partners pllc

Swaback Partners pllc

DC Ranch
Scottsdale, Arizona

The master plan for this 8,300-acre development is the fortunate outcome of a process marked earlier by conflict. At one point a legal dispute between the property owner and the city was settled by the Arizona Supreme Court with the imposition of a court-mandated plan that would have damaged both the environment and the resulting development. Five years later, the process was redone, with the developer and city collaborating on a plan that respects the land and assures a cohesive community through comprehensive architectural controls and governance. A mix of residential types is located within walking distance of a range of services and amenities, all of them linked by a system of paths and trails. The open space designed as part of the development combined with land set aside for city ownership amounts to 74 percent of the total acreage. Of the remaining 26 percent, about half is in private gardens and other non-building uses. Architectural guidelines recognize the need for texture, shade, and shelter in the desert environment. The siting of all buildings must be topography-sensitive, all plantings drought-tolerant, and watercourses treated as amenities. Public elements such as streets, bridges, lighting, signs, and fences must contribute to a coherent environment.

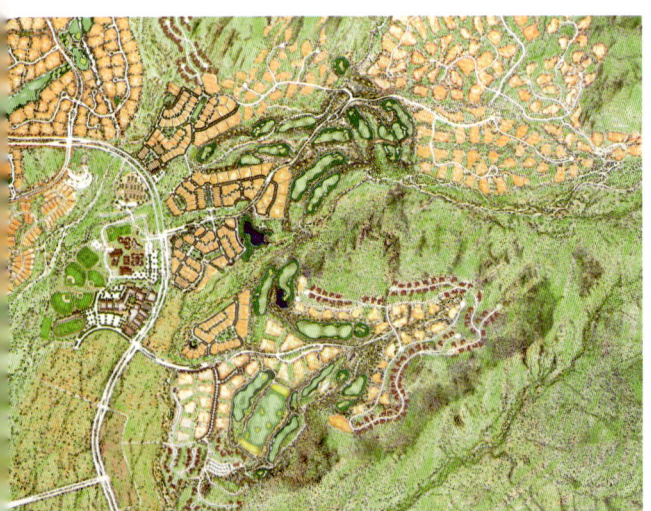

Above left: Partial master plan.

Left: Entry marker.

Top right: Portion of a trail system.

Above: Architecture combining openness with durable shelter.

Photography: DMB.

Top left: Golf course.

Above left: Sculpture in rock garden.

Left: Retail structure with discreet signage.

Right, top to bottom: View to city lights; community event; swimming oasis; clubhouse.

283

Swaback Partners pllc

Biosphere 2
Oracle, Arizona

Top of page: Plan for local mixed-use alternative.

Above left: Biosphere 2 structures.

Left: Possible hotel uses inside Biosphere.

Above: Renderings of possible gateway design and resort pools.

In January 2005 Swaback Partners was asked by developers to draw up a series of conceptual planning alternatives for the future development of an area centered on the world-renowned Biosphere 2. The land consists of the 140-acre Biosphere campus and two adjacent ranch tracts, which total 1,660 acres, and it offers striking panoramic views of the Santa Catalina Mountains. Alternatives proposed include: a hospitality complex, including resort hotel, health spa, retreat, and conference facilities; a cultural center, including performing arts, library, museum, and botanical garden; a research park, including a university annex and corporate facilities; a recreation complex, including an Olympic training center, an extreme sports center, and an ice rink; a local mixed-use development, including a town center, playground, amphitheater, and public plaza. Plans for the 1,520 acres that would surround any of these alternatives include a residential component of 1,553 homes, with 1,800 more approved for an adjacent 2,203 acres of State Trust land.

Swaback Partners pllc

Scottsdale Hangar One
Scottsdale, Arizona

Top: Entry courtyard, showing signature rooftop "paper airplane."

Top right: Canopied entrance near apron.

Above: Car showroom.

Above right: Canopies cantilevered from hangars.

Left: Entry past distinctive steel-tube structural grid.

Photography: Paul Warchol.

This 129,000-square-foot facility combines airplane hangars, offices, car showroom/storage, and aviation/automotive repair shops. It can accommodate up to 15 aircraft in its two 30,000-square-foot hangars. An approximately 68,500-square-foot multipurpose facility adjoining the hangars includes premium office space, entertainment space, and below-grade parking, as well as a unique high-end automobile showroom. The unique mix of uses has caused the project to be dubbed "a spa for jets." Hovering over the roof, a 108-foot-long aluminum "paper airplane" symbolizes the complex and is one of several exceptional structural accomplishments visible throughout the complex.

Swaback Partners pllc

The Village of Kohler
Kohler, Wisconsin

Swaback Partners has been involved for more than 25 years in the master planning and urban design of Kohler, a community originally planned in the early 1900s by the renowned Olmsted firm. A master plan drawn up in 1978 was updated in 1980 and again in 1990. The plan includes preservation of an 800-acre riparian open space zone and trail system, an astrophysical observatory, and golf course communities. In the early 1970s, the center of Kohler was being vacated as retail activity moved to a nearby mall. Now a village center provides a full range of goods and services, along with a lakeside inn. The American Club, a structure built in 1918 to house immigrant workers, has been restored as a highly-rated resort. Two 18-hole golf courses, created under the evolving master plans, are internationally known.

Left, top to bottom: Partial aerial view; open space along river; winter activity.

Above: Community event along Kohler's central lake.

Below: Golf course along Lake Michigan.

Bottom: American Club resort.

Photography: Kohler Co.

Swaback Partners pllc

Las Palomas
Puerto Penasco, Mexico

Left: Master plan.

Below left: Courtyard between high-rises.

Bottom left: Retail area at the base of towers.

Right: High-rise cluster seen from ocean.

Below right: Pool overlooking sea.

The master plan design placed high-rise residential complexes to take full advantage of ocean views. Amenities include golf, recreation, dining, lounging, and a variety of residential types. The clusters of towers define intimate pool and patio spaces. The architecture of the development includes contemporary design elements consistent with high-rise construction and modern technology, as well as traditional materials and forms that create a "village" experience, with narrow brick-paved passages, thick wall masses, and trellises as shading elements. The use of regional landscape materials and native stone, with work by local artists and craftsmen, contributes to a sense of timelessness and a connection to the place.

Swaback Partners pllc

Marana Master Plan
Marana, Arizona

Traditionally a small farming community near Tucson, Marana has grown to a population of 30,000 over the past decade, and demographic projections indicate an increase of 100,000 over the next 25 years. Swaback Partners has been retained to plan a 100-acre mixed-use town center that will establish a new identity for Marana and act as a guide for the character of future development. The planned urban village core will encourage higher-density housing and provide for public open space and industrial, office, and retail uses, along with civic and governmental services. The firm is also assisting the town in preparing residential design standards and overall land use scenarios for over 38,000 acres of undeveloped land in its total of 76,000 acres.

Top drawings, clockwise from top left: Master plan; overall view of town center; two views of proposed pedestrian environment.

Left: Aerial view of government buildings.

Thomas Balsley Associates

31 West 27 Street
New York, NY 10001
212.684.9230
212.684.9232 (Fax)
info@tbany.com
www.tbany.com

Thomas Balsley Associates

Thomas Balsley Associates

J-City
Tokyo, Japan

Two public spaces with Thomas Balsley sculptures provide two distinctive entrance environments and help to define the image for this new mixed-use project. Circulation paths are subtly indicated by a variety of elements that enliven these spaces for both passersby and building tenants. The main entrance plaza features a sculptural landform whose sloping stainless steel fountain and crescent shape attract visitors' attention to the front doors. This plaza is embellished with curving Corten sculptures and graphic walls whose cues have been taken from the book publishing heritage the neighborhood once nurtured. Retail cafes help activate this space throughout the week and the weekend. The project's residential units and conference rooms are entered through a terrace environment with lush gardens, fountains, and pergolas.

Above left: Stainless steel fountain sculpture where entry plaza meets public sidewalk.

Left: Graphic walls.

Below left: Sculpture with typographic motifs.

Above: Main plaza from above.

Left: Evening view from street.

Photography: Thomas Balsley (above, facing page top and center), Kokyu Miwa (all other).

Thomas Balsley Associates

Capitol Plaza
New York, New York

Left: Terrace with evening light.

Facing page, top left: Plaza passing through block.

Facing page, top right: Socializing at picnic tables.

Photography: Thomas Balsley (below and facing page below), Michael Koontz (all other).

The rezoning of the Sixth Avenue corridor in New York's Chelsea district provided for a through-block plaza whose success as a public space depends on sustained activity along its edges – similar to the traffic on a sidewalk. One edge of the plaza benefits from the new building's cafes and entrances, but the other edge required a different strategy. What was originally a 200-foot-long blank wall is now a new urban edge, carved out both as a narrow retail storefront and as a bright orange wall with oval penetrations, through which bamboo emerges. The plaza's contemporary design sensibility is a purposeful outreach to the neighborhood's creative professionals who work long studio hours and shoppers who flock to the neighborhood's unique retail "bazaar" environment. A wide range of seating and socializing choices, ranging from raised bar tables with swivel stools and benches with laptop tablettes to urban picnic tables, attracts working and conferencing well beyond the normal lunch peak periods.

Left: Boldly colored wall commanding attention from street.

Right: Raised terrace in bamboo glade.

Thomas Balsley Associates

Pacific Design Center
West Hollywood, California

Left: Fountains and gardens in front of the "Green" building, an annex to original "Blue Whale."

Below: New "Wave Park" in front of Blue Whale.

Below left: Three detail view of grassy amphitheater and undulating lawn panels.

Photography: Tom Hinckley (left and under), Jay Venezia (all other).

The five-acre setting of this Modernist landmark by architect Cesar Pelli – its main building known affectionately as the "Blue Whale" – has been transformed into a lively park-like landscape supporting the repositioning of the showroom complex as a multipurpose destination. Once a barren plaza, the space along Melrose Avenue, renamed "Wave Park" for it undulating panels of lawn, invites a variety of lounging and picnicking experiences. A destination café kiosk and terrace complement the new entrance and automobile arrival area. Red walls with recessed cobalt blue lights frame each lawn panel. A fountain plaza with terraced lawns and gardens has opened views and public access to the center. With the new building illumination and graphic sculptures, they create evening drama and a new public image for the facility.

Thomas Balsley Associates

Riverside Park South
New York, New York

Above: Detail view of pier.

Right, top to bottom: Coastal grasses, with old railroad gantry in background; overlook terrace; aerial view of pier, cove, and gantry; boardwalks through grasses.

Photography: Michael Koontz.

The new urban development from 59th to 72nd street will include a new 27-acre waterfront park, giving westsiders direct access to their river for the first time. The first phase of the park, three acres along the Hudson River stretching from 65th to 70th Streets, provides a series of river-edge experiences involving an extensive network of boardwalks, overlooks, and plaza spaces at the ends of the cross streets. Riprap is used at the water's edge to blur the lines between river-edge grasses, public lawns, and the Hudson. This softer river edge, along with the low elevations of the boardwalks, offers pedestrians more opportunity to get close to the water. Standing in dramatic contrast to these elements, an old railroad gantry tower has been stabilized and left in place. Dedicated bike and in-line skating paths connect Phase One of the park to the bike path system to the south. Elevated above the water's edge is a series of timber and steel overlooks providing viewing places and seating areas along the main walkway. Following the angled lines of the former railroad piers, the overlooks include interpretive signs telling the history of the site.

Thomas Balsley Associates   World Trade Center Plaza
Osaka, Japan

Conceived as a water garden environment, this public space serves the Center as its front door as well as an outdoor extension of the winter garden's atrium and restaurants. Its terraces are surrounded by a lush landscape and a water feature whose raised basin is aligned with the visitor's sitting eye level and releases waves down its crescent slope. On the water basin's street edge is a water slide whose descent is interrupted by a grid of metal pipe protrusions. The plaza's dominant feature, the set of cone sculptures, has become the Center's iconic element. At first glance, they appear to serve only that purpose, but just beneath the surface, literally, are the mechanical rooms whose need for fresh air intake is accommodated by the cones' perforations. Emitting light and mist in the evening, these sculptures enrich the experience for the Center's workers and visitors.

Top: Aerial view.

Above left: Pool and cones at night.

Above: Plaza with building entrance.

Far left: Paths extending plaza experience.

Left: Raised basin and outdoor dining area.

Photography: Cervin Robinson.

Thomas P. Cox: Architects, Inc.

19782 MacArthur Boulevard
Suite 300
Irvine, CA 92612
949.862.0270
949.862.0289 (Fax)
info.oc@tca-arch.com
www.tca-arch.com

600 Wilshire Boulevard
Suite 1470
Los Angeles, CA 90017
213.553.1100
213.553.1111 (Fax)
info.la@tca-arch.com

Thomas P. Cox: Architects, Inc.

Botanica on the Green
Stapleton, Colorado

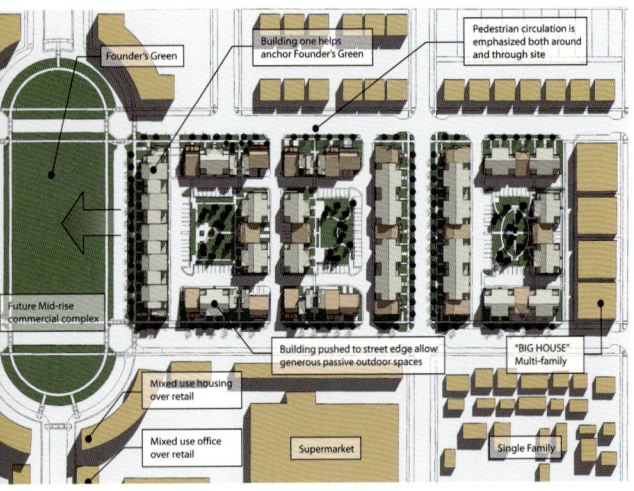

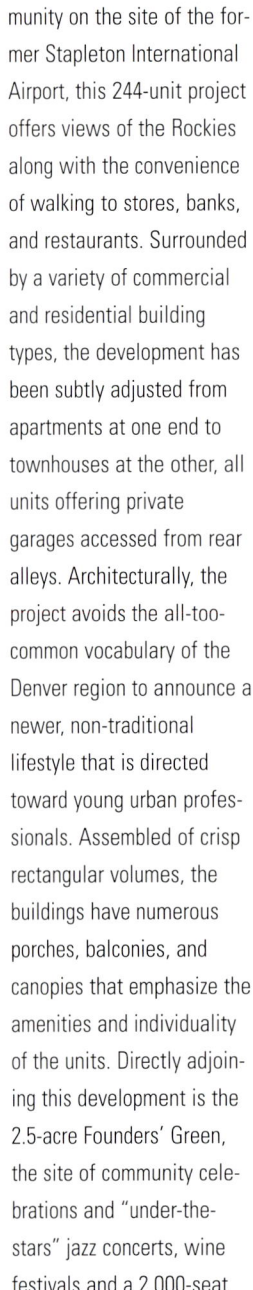

As part of a planned community on the site of the former Stapleton International Airport, this 244-unit project offers views of the Rockies along with the convenience of walking to stores, banks, and restaurants. Surrounded by a variety of commercial and residential building types, the development has been subtly adjusted from apartments at one end to townhouses at the other, all units offering private garages accessed from rear alleys. Architecturally, the project avoids the all-too-common vocabulary of the Denver region to announce a newer, non-traditional lifestyle that is directed toward young urban professionals. Assembled of crisp rectangular volumes, the buildings have numerous porches, balconies, and canopies that emphasize the amenities and individuality of the units. Directly adjoining this development is the 2.5-acre Founders' Green, the site of community celebrations and "under-the-stars" jazz concerts, wine festivals and a 2,000-seat amphitheater.

Left: Site plan, with Founders' Green at left.

Below left: Apartment units seen from Green.

Bottom left: Sheltered entries, porches, and balconies of townhouses.

Right: Row of apartment homes.

Opposite bottom: Row of townhouses.

Photography: Steve Hinds Photography, Dallas TX.

Thomas P. Cox: Architects, Inc.

Crescent Park Apartment Homes
Playa Vista, California

Left: Site and first residential level plan.

Below: Main front on North Crescent East, with pool court at center.

Below right: Portion of Playa Vista Drive front, with garden court at left.

Bottom right: Interior of leasing office.

Facing page: Exterior detail, with main entrance and pool court trellis.

Photography: Steve Hinds Photography, Dallas TX.

Located in Playa Vista, Los Angeles' first master-planned community in over 50 years, this 214-unit development was subject to many challenges in terms of zoning and design controls. Architectural styles were "dealt" to each developer to ensure design diversity in the community. This project was assigned the style of the pioneering California Modernist Irving Gill, whose work was notable for its smooth surfaces, sharp edges, and round arches. But while Gill's style exploited the characteristics of tilt-up concrete construction, this project had to recall his forms with wood and stucco. And since Gill never worked at the scale of this project, the building forms had to be broken down visually with advancing and receding planes and changes in color. The project includes four levels of apartments ranging from one-bedroom to two-bedrooms-plus den, notable for their 9-to-11-foot ceilings and high standards of fixtures and appliances. Shared amenities include a fitness center, spa and a business center with conference room. The distinctive garden courts and pool court provide views for residents and divide the complex into smaller-scaled volumes.

Thomas P. Cox: Architects, Inc.

Grand Avenue Competition
Los Angeles, California

TCA was a key participant in a runner-up team for the multibillion-dollar Grand Avenue Project, a redevelopment effort that is expected to link the civic and cultural districts of Los Angeles in a vibrant new regional center. The national competition for the project was open to any developer capable of proposing a 3.2-million-square-foot mini-city programmed to include entertainment venues, restaurants, retail, office buildings, a hotel, and new housing. TCA was a member of the Forest City Enterprises' team, which also included A.C. Martin & Partners and Calthorpe Associates. Housing was TCA's major contribution of the team's proposal, which promised 4,700 units, including several high-rise apartment towers. A 16-acre "Great Park," stretching three city blocks from the City Hall to the cultural precinct along Grand Avenue, would have redeveloped an existing underused open mall, which is now hidden by an old county building. The project as a whole is expected to generate 16,000 permanent jobs and contribute $85 million annually in taxes. TCA is making other contributions to the revitalization of downtown L.A. with the renovation of the historic Subway Terminal Building, for Forest City Enterprises, and an 850-unit high-rise mixed-use project for the Little Tokyo area for the Related Companies, the developer that won the Grand Avenue competition.

Left and top left: Residential towers with mixed-use bases.

Far left, top to bottom: View of "Great Park," with City Hall tower in foreground; Grand Avenue, with Chandler Pavilion in foreground, downtown skyline beyond; View of "Great Park," with City Hall in background.

Facing page: Plans and aerial visualization of competition proposal.

Thomas P. Cox: Architects, Inc.

Westgate – Pasadena
Pasadena, California

Left: Project plan in context of downtown.

Below: Entrance to pedestrian passage.

Bottom left: Pedestrian street with variety of housing types.

Bottom right: Residential units above street-level retail.

Development of this 12-acre site, adjoining the historic core of Pasadena, will reinforce the ambitious downtown revitalization effort. Today, the area consists mainly of abandoned industrial facilities and parking lots, acting as a barrier to pedestrian access to downtown from residential areas beyond. The design of the neighborhood includes 832 residential units, for both rental and for-sale, in a variety of 3-to-5-story configurations, with unit front doors typically off public streets or open spaces. Parallel parking on streets satisfies a fraction of the parking need, with the rest beneath the buildings. Pedestrian movement is encouraged by a system of paseos and courtyards penetrating the blocks. Retail components are included adjacent to existing downtown retail. This project has already been recognized by The Congress for the New Urbanism in 2005 when it received an Award of Excellence for The Neighborhood, the District, and the Corridor, and the American Institute of Architects, Orange County in 2004 for an Honor Award in Planning.

PROJECT CREDITS

Exceptional care has been taken to gather information from firms represented in this book and transcribe it accurately. The publisher assumes no liability for errors or omissions in the credits listed below.

180° DESIGN STUDIO

New Town Theater District
Client: Whittaker Builders, Inc.
Principal Consultants: 180° Design Studio, urban design, architectural design
Pickett Ray & Silver, engineers

New Longview
Client: Gale Communities, Inc.
Principal Consultants: 180° Design Studio, master plan, design charrette, architectural design (residential)
Patti Banks Associates, landscape
Zimmerman Volk Associates, housing market
Gibbs Planning Group, retail market
Swift & Associates, engineering design
Hamilton, Sterrett & Dooley, civil engineering
Allison Ramsey Architects, architectural design
Patrick Pinnell, architectural design

Crescent Creek
Client: Dial Realty
Principal Consultants: 180° Design Studio, master plan, architectural design
Missouri Valley Engineering, engineering
Patti Banks Associates, landscaping

Ottawa University Master Plan
Client: Ottawa University
Principal Consultants: 180° Design Studio, master planning

71st and Metcalf
Client: Dial Realty
Principal Consultants: 180° Design Studio, master planning

Union Hill Phase 3
Client: Union Hill Phase 3
Principal Consultants: 180° Design Studio, urban design, architectural design
Patti Banks Associates, landscape
Sullivan/Palmer, construction drawings

Beachtown Galveston Model Home
Client: Beachtown Galveston Village
Principal Consultants: 180° Design Studio, master planning, architectural design

Longfellow Court
Client: 180° Development, LLC
Principal Consultants: 180° Design Studio, master planning, architectural design

ALTOON + PORTER ARCHITECTS

Knox Shopping Centre
Client: AMP Henderson Global Investors
Principal Consultants: Altoon + Porter Architects, architectural design consultant
Hames Sharley International Ltd., associated architect
Bonacci Winward, structural/civil engineer
Simpson Kotzman Pty. Ltd., mechanical/electrical engineer
Tract, landscape architect
ARUP Fire, fire engineer
NDY Light, specialty lighting
Rawlinsons Pty. Ltd., quantity surveyor
Grogan Richards Pty. Ltd., traffic

C.J. Arms & Associates, hydraulics
Transportation Design Consultants Pty. Ltd., vertical circulation
ProBuild, general contractor

Victoria Gardens
Client: Forest City Development California, Inc.
Principal Consultants: Altoon + Porter Architects, architectural design
KA Inc. Architecture, executive architect
Field Paoli Architects, design architects
Elkus/Manfredi Architects Ltd, design architects
Thorson Baker & Associates, Inc., structural engineer
S.Y. Lee Associates, mechanical/plumbing engineer
Patrick Byrne & Associates, electrical engineer
MDS Consulting, civil engineer
SWA Group, landscape architect
Kaplan Partners Architectural Lighting, lighting designer
Redmond Schwartz Mark Design, environmental graphics
The Mobility Group, traffic engineer
Vratsinas Construction Company (VCC), construction manager
Forest City Construction Co., Inc., construction

Fashion Show
Client: The Rouse Company (project currently owned by General Growth Properties)
Principal Consultants: Altoon + Porter Architects, architectural design
Richard Orne, AIA, Orne & Associates, and Laurin B. Askew, Jr., FAIA, Monk LLC, conceptual design
Richard Orne, AIA, Orne & Associates, design manager for The Rouse Company
Ove Arup & Partners; ASI; SME Steel Contractors: structural
Tschuchiyama & Kaino, mechanical/plumbing
Patrick Byrne & Associates, electrical engineer
B&R Construction Services, electrical
G.C. Wallace, Inc., civil engineer
SWA Group, landscape architect
Rolf Jensen & Associates, Inc., fire & life safety code
RWDI, cloud canopy
Terracon Consultants Western, Inc., geotechnical engineer
Zipper Zeman Associates, Inc., geotechnical/environmental
Chew Specifications, specifications
Stantec Consulting, ADA
Shen, Milson, Wilke-Paoletti, Inc., acoustician
Enterscapes Entertainment, entertainment systems
Kaplan Partner Architectural Lighting; Lightswitch, decorative lighting
Gallegos Lighting Design, theatrical lighting
Central Parking System; Walker Parking, parking
Gorove/Slade Associates, Inc., traffic engineer
Sussman/Prejza Company, Inc., graphics/signage
Audio Visual (Structural); Vantage Technology Consulting Group, technology consultant
Ford Audio Video, audiovisual investigator
CM Resources, Inc., audiovisual consulting
Mendenhall Moreno & Associatesl, Inc., zoning
The Whiting-Turner Contracting Company, general contractor

ANNEX/5

Motorola Global Software Group
Client: Motorola
Principal Consultants: Annex/5, site master planning, architectural design and documentation
Epstein Sp. z.o.o. structural and mep engineering

Planning Design of Haikou West Coast
Client: Planning Bureau of Haikou City, Haikou Municipal Government
Principal Consultants: Annex/5, detailed master plan and design guidelines
Epstein International (China) Co., Ltd.

Brighton Village
Client: Pyramid Group
Principal Consultants: Annex /5, master planning and architectural design of town center
CMX, civil engineer
Design Workshop (not shown), subdivision master planner

Serta International Center
Client: Serta, Inc.
Principal Consultants: Annex/5, site master planning, architectural design, construction documents
Epstein Engineering, structural, civil, mep
Jacobs/Ryan, landscape architect

AUSTIN VEUM ROBBINS PARTNERS

Allegro Tower Apartments
Client: Leo Frey
Principal Consultants: Austin Veum Robbins Partners, architectural design
Glotman Simpson, structural
McParlane and Associates, mechanical and plumbing
ILA Zammit, electrical
KTU&A, landscape
DPR Construction, general contractor

Egyptian Lofts
Client: Citymark Egyptian LLC
Principal Consultants: Consultants: Austin Veum Robbins Partners, construction administration, architectural design
Deneen Powell Atelier, landscape
Glotman Simpson, structural
Project Design Consultants, civil
McParlane and Associates, mechanical
Neal Electric, electrical
Megan Bryan Studio, interiors
Swinerton Builders, general contractor

Park Laurel on the Prado
Client: CLB Partners
Principal Consultants: Austin Veum Robbins Partners, full architectural services
Magnusson Klemencic & Associates, structural
Roel Construction, contractor

Smart Corner
Client: Smart Corner, LLC
Principal Consultants: Austin Veum Robbins Partners, full architectural services, site selection and feasibility
Davis Davis Architects, office building architects
Magnusson Klemencic & Associates, structural

McParlane & Associates, mechanical and plumbing
Kruse & Associates, electrical
Project Design Consultants, civil/survey/landscape
Hensel Phelps Construction Co., contractor

The Pinnacle Museum Tower
Client: Pinnacle International
Principal Consultants: Austin Veum Robbins Partners, full architectural services
Sterling Cooper & Associates, mechanical engineers
Glotman Simpson, structural engineers
Nemetz & Associates, electrical engineers
Schirmer Engineering, building code and fire safety consultants
P&D Consultants, civil engineers and surveyors

BEELER GUEST OWENS ARCHITECTS, L.P.

The Highlands of Lombard
Client: Lincoln Property Company
Principal Consultants: Beeler Guest Owens Architects, full architectural services
Virgilio & Associates, LTD, structural engineer
Lehman Design Consultants, mep engineer
V3 Companies, civil engineers,
Enviro Design, landscape architects

5225 Maple Avenue
Client: MAEDC
Principal Consultants: Beeler Guest Owens Architects, full architectural services
Brockette Davis Drake, Inc., civil engineer
Enviro Design, landscape architect
Brockette Davis Drake, Inc., structural engineer
BEI Basharkhah Engineering, Inc., mep engineer

Easton
Client: Phoenix Property Company
Principal Consultants: Beeler Guest Owens Architects, full architectural services
Jerald W. Kunkel Engineering, structural engineer
BEI Engineering, Inc., mep engineer
Enviro Design, landscape architects
Ferguson-Deere, Inc., civil engineer
North American Precast Company, garage consultant

The Davis Building
Client: Hamilton & Davis, L.L.P.
Principal Consultants: Beeler Guest Owens Architects, full architectural services
Parkin-Perkins-Olson, Inc., structural engineer
S Toab & Associates, mep engineer
J.A. Jones Construction, contractor

Union Station
Client: TKV Union Station, L.P.
Principal Consultants: Beeler Guest Owens Architects, full architectural services
Gannett Fleming, landscape architect
The Reynolds Group, Inc., civil engineer
O'Donnell & Naccarato, garage/foundation engineer
Sterling Engineering and Design Group, LTD., framing engineer
Schoor Depalma, traffic engineer
Basharkhah Engineering, Inc., mep engineer
Kathy Andrews Interiors, interior designer
Metropolitan Acoustics, LLC., acoustical consultant

CALLISON ARCHITECTURE, INC.

Ayala Center Greenbelt
Client: Ayala Land, Inc.
Principal Consultants: Callison Architecture, Inc., master planning, architecture, interior design, merchandise strategy, tenant strategy, environmental graphic design
GF & Partners, associate architects
Edward D. Stone & Associates, landscape design
ACL Asia, landscape design
Rey J. Calpo & Partners, mechanical engineer
DCCD Engineering Corp., electrical engineer
NBF Consulting Engineers, sanitary engineer and fire protection design
Davis, Langdon & Seah Philippines, quantity surveyor
Makati Development Corp., general contractor

Grand Gateway
Client: Hang Lung Development Co., Ltd.
Principal Consultants: Callison Architecture, Inc., master planning, architecture, interior design, environmental graphic design, leasing strategy, tenant criteria
Stamper Whitin Works, landscape architect
Horton Lees Lighting Design, Inc., lighting design
Frank C.Y. Feng Architects and Associates, Ltd., associate architect
Maunsell Consultants Asia, Ltd., civil/structural
Associated Consulting Engineers, mechanical/electrical
Levett & Baileyl, quantity surveyor
Fujita Corporation, contractor

Suwon Gateway Plaza
Client: Suwon Aekyung Station Development Co., Ltd.
Principal Consultants: Callison Architecture, Inc., master planning, architecture
Suh-han Architects & Engineering Inc., local architect of record
Magnusson Klemencic Associates, structural engineer
Hargis Engineering, Inc., lighting designer
Sean O'Connor Associates, lighting designer

Metropolitan Tower
Client: Continental Bentall, LLC
Principal Consultants: Callison Architecture, Inc., architecture
Cary Kopczynski & Company, structural engineering
Carter Burgess, mechanical engineering
KPFF, civil engineering
Sechrist Design Associates, interior design
Heier Design Group, landscape architecture
Mortensen Construction, general contractor

CANIN ASSOCIATES

Grande Lakes Resort
Client: Marriott International
Principal Consultants: Canin Associates, development approval, master planning in partnership with Site Concepts International
Site Concepts International, landscape architecture
Smallwood Reynolds Stewart Stewart, architects
Donald W. McIntosh Associates, Inc., civil engineers

Buena Vista
Client: Castle & Cooke Inc.
Principal Consultants: Canin Associates, master planning, landscape architecture, residential design
McIntosh & Associates, Inc., civil engineers

Solivita
Client: Avatar Properties Inc.
Principal Consultants: Canin Associates, entitlement acquisition, master planning, landscape architecture
Spillis Candela & Partners, Inc., architects

Promenade Town Center
Client: Crown Community Development
Lead Consultant: Canin Associates, master planning
URS Corporation, commercial development

CARTER & BURGESS, INC.,

North Hills
Client: Kane Realty Corporation
Principal Consultants: Carter & Burgess, Inc., master planning, architectural design, architect of record, mep, structural.
Mahan Ryhiel & Associates, Inc., landscape architect
The John R. McAdams Co., Inc., civil engineer
The Lighting Practice, lighting designer

Tha Walk
Client: Casino Reinvestment Development Authority (CRDA), Owner
The Cordish Company, Developer
Principal Consultants: Carter & Burgess, Inc., master planning, architectural design, environmental graphics, architect of record, mep, structural.
Pennoni Associates, Inc., civil engineer
Greenald-Waldron Associates, lighting consultant
J. Adamson Associates, landscape architect

Power Plant Live!
Client: The Cordish Company
Principal Consultants: Carter & Burgess, Inc., planning, program and environmental graphics design, project theme, tenant primary Identification, lighting, softscape, hardscape, wayfinding, project identification, furniture plan and plaza bar design.
Brown & Craig, Inc., architect of record
Stern & Associates, Inc., mep
Stone Mountain Lighting, lighting consultant
Century Engineering, structural
Hope Furrer Associates, structural

Citrus Plaza
Client: Majestic Realty Co.
Principal Consultants: Carter & Burgess, Inc., master planning, architectural design, environmental graphics, architect of record.
Ajit S. Randhava & Associates, structural
Environs, landscape architect
Commerce Construction LP, mep, design build

The Shoppes at Blackstone Valley
Client: WS Development/SR Weiner
Principal Consultants: Carter & Burgess, Inc., master planning, architectural design, environmental graphics, architect of record, mep, structural, landscape design.

Walkers Brook Crossing
Jordan's Furniture, Home Depot, Imax
Client: Dickinson Development, Developer
Principal Consultants: Carter & Burgess, Inc., master planning, architectural design, environmental graphics, architect of record, mep, structural.
VHB, civil engineer
Haley and Aldrich, geotech
Ernest Gould, site lighting

CHAN KRIEGER & ASSOCIATES

Beth Israel Deaconess Medical Center & Master Plan
Client: Beth Israel Deaconess Medical Center
Principal Consultants: Chan Krieger & Associates, architectural design, urban design, streetscape design
Rothman Partners, Inc., architect-of-record
Solomon + Bauer, interior architecture
Childs Associates, landscape architecture

City Hall Plaza Community Arcade and Government Center Master Plan
Client: City of Boston/Boston Redevelopment Authority, Trust for City Hall

Principal Consultants: Chan Krieger & Associates, master planning, urban design, architecture
LeMessurier Associates, structural engineering
LAM Partners, lighting design

Three Rivers Park
Client: RiverLife Task Force, Forest City Commercial Group
Principal Consultants: Chan Krieger & Associates, urban design, architectural design
Hargreaves Associates, landscape
Bohlin Cywinski Jackson, architecture and planning
Urban Instruments, architecture and public art
Economic Research Associates, economic development
Parsons Brinckerhoff Quade & Douglas, transportation and environmental planning

Fort Washington Way Highway Reconfiguration
Client: City of Cincinnati
Principal Consultants: Chan Krieger & Associates, urban and architectural design services
Parsons Brinckerhoff Quade & Douglas, lead engineer/project management
KZF Design, project architects
Hargreaves Associates, landscape architects

CHARLAN • BROCK & ASSOCIATES, INC.

Williams Walk at Bartram Park
Client: Daniel Corporation
Principal Consultants: Charlan • Brock & Associates, Inc., architectural design
KTD Engineers, mep
N.H. Joshi & Associates, structural engineer
Dix-Lathrope, landscape and land planning
Carnegis & Co., interiors

Aqua Condominiums
Client: Aronov Realty Management, Inc.
Principal Consultants: Charlan • Brock & Associates, Inc., architectural design and land planning
McNeil Engineering, Inc., civil engineer
Dix-Lathrope Landscape Architects, landscape architect
N.H. Joshi & Associates, structural engineer
KTD Engineers, mep
Florida Coastal Development Consulting, Inc., pool engineer
W.G. Yates Construction Co., general contractor

Uptown Maitland West
Client: Uptown Maitland West
Principal Consultants: Charlan • Brock & Associates, Inc., architectural design and land planning
Dyer Riddle Mills Precourt, civil engineer
Dix-Lathrope Landscape Architects, landscape architect
N.H. Joshi & Associates, structural engineer
KTD Engineers, mep
Kelsey Construction, general contractor

Rarity Pointe Lodge and Spa
Client: RPL Properties Inc.
Principal Consultants: Charlan • Brock & Associates, Inc., architectural design and land planning
Sterling Engineering, Inc., civil engineer
Hawkins Partners, Inc., landscape architect
N. H. Joshi & Associates, structural engineer
KTD Engineering Consultants, mep
Godfrey Design Consultants, Inc., interior design
Raleigh Design, interior design
Hardin Construction Company, general contractor
Lawler-Wood, LLC, developer

The Flats at Rosemary Beach
Client: Lowder Construction
Principal Consultants: Charlan • Brock & Associates, Inc., architectural design

Connely & Wicker, civil engineer
N. H. Joshi & Associates, structural engineer
KTD Engineering Consultants, mep
Lovelace Interiors Inc., interiors
Lowder Construction, general contractor

Cheval Apartments on Old Katy Road
Client: The Spanos Corporation
Principal Consultants: Charlan • Brock & Associates, Inc., architectural design and land planning
Carter-Burgess, civil engineer
Vernon G. Henry and Associates, Inc., land planner
K & W (Kudela & Weinheimer), landscape architect
N.H. Joshi & Associates, structural engineer
E.N., Inc., mep

CBT/CHILDS BERTMAN TSECKARES, INC.

Columbus Center
Client: Cassin-Winn Development Company
Principal Consultants: CBT/Childs Bertman Tseckares Inc., architects
Cosentini Associates, Consulting Engineers (mep, lighting, security, tele/data)
Harry R. Feldman, Inc., land survey
Haley & Aldrich, geotechnical
Judith Nitsch Engineering, Inc., traffic
BBG-BBGM, hotel
Central/Myers, parking
Pressley Associates, Inc., landscape architects
LeMessurier Consultants, structural engineering
Hughes Associates, Inc., code/life safety
Edgett Williams Consulting Group Inc., elevator
Cerami Associates, acoustical
Campbell-McCabe Inc., hardware
Zaldastani Associates, Inc., tunnel structure
Hatch Mott McDonald, tunnel engineering

North Point
Client: Spaulding & Slye Colliers & Guilford Transportation Industries, Developer/Development Manager
Principal Consultants: Kenneth Greenberg, Greenberg Consultants Inc., in association with CBT/Childs Bertman Tseckares, Inc., master planners
CBT/Childs Bertman Tseckares, Inc., architect, phase I team, parcel S residential building
Architects Alliance, architect, parcel T residential building
Michael van Valkenberg Associates, landscape architect, central park

The Residences at Kendall Square
Client: TP Kendall LLC
Principal Consultants: CBT/Childs Bertman Tseckares, Inc., architects
Cosentini Associates, consulting engineers
Bovis Lend Lease, general contractor

Rollins Square
Client: Planning Office for Urban Affairs / Archdiocese of Boston
Principal Consultants: CBT/Childs Bertman Tseckares, Inc., Architects
CBA Landscape Architects, landscape architect
Weidlinger Associates, structural engineer
Fitzemeyer & Tocci, mechanical engineer
Peter J. Roche, Real Estate and Community Development, development consultant
Suffolk Construction, general contractor

The Prudential Center Redevelopment
Client: Boston Properties
Principal Consultants: CBT/Childs Bertman Tseckares, Inc., architects
TMP Consulting Engineers, mep engineer

McNamara/Salvia, Inc., structural engineer
Vanasse Hangen Brustlin, Inc., civil & traffic engineer
Carr, Lynch and Sandell, landscape architect
Gordon H. Smith Corporation, exterior wall
LAM Partners Inc., lighting
Rolf Jensen & Associates, code
Cavanaugh Tocci Associates, Inc., acoustics
RTE Group, Inc., teleData
Selbert Perkins Design Collaborative, graphic design
Haley & Aldrich, Inc., soils engineer (geotechnical)
John Van Stone Fogg, CSI, specifications

COSTAS KONDYLIS AND PARTNERS LLP

Trump World Tower – 845 United Nations Plaza
Client: Daewoo-Trump
Principal Consultants: Costas Kondylis and Partners LLP, architectural design
Israel Berger & Associates, structural engineer
I.M. Robbins, mechanical engineer
Gordon H. Smith Company, curtain wall consultants
Brennan Beer Gorman/Monk Interiors, interior design (lobby)
Abel, Bainnson & Butz, landscape architect

The Heritage
Client: Hudson Waterfront Associates/Trump New World Project Management
Principal Consultants: Costas Kondylis and Partners LLP, architectural design
Rosenwasser/Grossman: structural engineer
I.M. Robbins, mechanical engineer
McGinley Design, interior design
Abel, Bainnson & Butz, landscape architect
HRH Construction, general contractor

Morton Square
Client: J.D. Carlisle Development Corp.
Principal Consultants: Costas Kondylis and Partners LLP, architectural design
Rosenwasser/Grossman: structural engineer
I.M. Robbins, mechanical engineer
Israel Berger & Associates, exterior wall consultants
Kondylis Design, interior design
Oehme, van Sweden & Associates, landscape architect
Philip Koether, interior design (lobby)
Tom Patti, glass artist

The Grand Tier
Client: Glenwood Management Corp.
Principal Consultants: Costas Kondylis and Partners LLP, architectural design
Rosenwasser/Grossman: structural engineer
I.M. Robbins, mechanical engineer
Saladino Group, Inc., interior design
Cosentini Associates, lighting design

CUNNINGHAM + QUILL ARCHITECTS PLLC

The Mather Building
Client: PN Hoffman Development
Principal Consultants: Cunningham + Quill Architects, architectural design
Ehlert/Bryan, Inc., structural engineers for senior center
Summit Engineers, Inc., mechanical, electrical and plumbing engineers for senior center
Best/Joslin, façade restoration consultant

Huntfield Master Plan
Client: Greenvest, LLC
Principal Consultants: : Cunningham + Quill Architects, master planning, urban design guidelines

Caton's Walk
Client: RB Associates
Principal Consultants: Cunningham + Quill Architects, architectural design
Ehlert/Bryan, Inc., structural engineers for senior center
Summit Engineers, Inc., mechanical, electrical and plumbing engineers for senior center

The Alta
Client: PN Hoffman Development
Principal Consultants: Cunningham + Quill Architects, architectural design
Ehlert/Bryan, Inc., structural engineers
GHT Limited, mechanical, electrical and plumbing engineers

National Cathedral School
Client: National Cathedral School
Principal Consultant: Cunningham + Quill Architects, master planning and site design

Park Hill Condominium
Client: PN Hoffman Development
Principal Consultants: Cunningham + Quill Architects, architectural design
Ehlert/Bryan, Inc., structural engineers
Bansal & Associates, Inc., mechanical, electrical and plumbing engineers

Fortnightly Neighborhood Master Plan and Herndon Senior Center
Client: Fairfax County Redevelopment Housing Authority Department of Housing and Community Development
Principal Consultants: Cunningham + Quill Architects, master planning, urban design Guidelines, Full Architectural Services for Senior Center
Oculus, landscape design for master plan and urban design guidelines
Ehlert/Bryan, Inc., structural engineers for senior center
Summit Engineers, Inc., mechanical, electrical and plumbing engineers for senior center
Riley & Rohrer, interior design services for senior center

DAHLIN GROUP ARCHITECTURE PLANNING

Black Diamond
Client: A. F. Evans
Principal Consultants: Dahlin Group Architecture Planning, urban planning, architectural design
Gates and Associates, landscape
Johnstone Moyer, construction manager
Main Street Property Services, retail consultant
Winzler & Kelly Consulting Engineers, civil engineer
Colour Studio, color consultant

Luxe Hills International Golf Community
Client: Wide Horizon Real Estate Development Company
Principal Consultants: Dahlin Group Architecture Planning, master planning, community building architecture, luxury estate villas
Nicolay Designs, landscape architects
JMP Golf Design Group, golf course architect

University Villages
Client: Marina Community Partners LLC
Principal Consultants: Dahlin Group Architecture Planning, master plan, urban design, residential architecture
MBH Architects, retail architects
RBF, civil engineer
The Guzzardo Partnership, landscape architect
Bob Schaffer, community outreach
Keyser Marston Associates, economic consultant
Zander & Associates, biology
Staub Forestry and Environmental Consulting, consulting forester and arborist

EIP, environmental consultant
McDonough, Holland & Allen PC, legal advisors
Lombardo & Gilles, legal advisors
Carol Lind, market research
Berlogar Geotechnical, soils/geologic engineer
Xenergy, sustainability

Coyote Valley
Client: City of San Jose
Principal Consultants: Dahlin Group Architecture Planning, lead consultant for plan, design guidelines
Ken Kay Associates, landscape architect, environmental footprint and landscape details
HMH Engineering, infrastructure design
EPS, economic consultant
Apex Strategies, community outreach

DAVID M. SCHWARZ

Ft. Worth
Sundance East
Client: Sundance East Partners
Principal Consultants: David M. Schwarz/Architectural Services, Inc., design architect
HKS, Inc., architect of record
Linbeck Construction Company, contractor

Ft. Worth
Sundance West
Client: Sundance West Partners
Principal Consultants: David M. Schwarz/Architectural Services, Inc., design architect
HKS, Inc., architect of record
Linbeck Construction Company, contractor

Ft. Worth
Nancy Lee and Perry R. Bass Performance Hall
Client: Performing Arts Fort Worth, Inc.
Principal Consultants: David M. Schwarz/Architectural Services, Inc., design architect
HKS, Inc., architect of record
Jaffe Holden Acoustics, acoustic/AV consultant
AltieriSeborWieber LLC Consulting Engineers, mep
Walter P. Moore & Associates, structural
Fisher Dachs Associates, theatre consultants
Linbeck Construction Company, contractor

Ft. Worth
Fort Worth Public Library
Client: City of Fort Worth
Principal Consultants: David M. Schwarz/Architectural Services, Inc., design architect:
Hidell Associates Architects (Interior), architect of record
Growald Architects (Exterior), architect of record
Huitt-Zollars, mep
Walter P. Moore & Associates, structural
Ratcliff Construction, contractor

Ft. Worth
Bank One Building
Client: Sundance Square Management, LLC
Principal Consultants: David M. Schwarz/Architectural Services, Inc., design architect
HKS, Inc., architect of record
Datum Engineering, structural
James Johnston & Associates, mep
Brockette Davis Drake, civil
FMG Design, Inc., graphics/signage
Linbeck Construction Company, contractor

Parker Square
Client: Five Star Development
Principal Consultants: David M. Schwarz/Architectural Services, Inc., design architect & master planner
Vidaud Associates (Bldgs. 8, 9), CSI (Bldgs. 3, 4), Five Star Development (Bldgs. 5, 6, 7), architects of record

Southlake Town Square
Client: Cooper and Stebbins
Principal Consultants: David M. Schwarz/Architectural Services, Inc., design architect & master planner
(Phase 1): Urban Architecture; subsequent phases: Urban Architecture*, Looney Ricks Kiss*, Bowie Griddley Architects*, Beck, architect of record:
* both Architect and Architect of Record responsibilities

West Village
Client: Phoenix Property Company and Urban Partners, Inc.
Principal Consultants: David M. Schwarz/Architectural Services, Inc., design architect & master planner
KSNG Architects, Inc., architect of record:
Dalmac Construction, contractor
L. A. Fuess Partners Engineers, structural

DOUGHERTY SCHROEDER & ASSOCIATES, INC.

Destin Commons
Client: Turnberry Associates, Legendary, Inc., and Retail Estate
Principal Consultants: Dougherty Schroeder & Associates, Inc. (full architectural services; architect of record)
Lee Richardson & Associates, landscape architect
Bliss Fasman, Inc., lighting design
Communication Arts, graphic design
Brady & Anglin Engineers, mep
Connelly & Wicker, Inc., civil engineer
Pruitt Eberly Stone, Inc., structural engineer
Tower Construction, general contractor

The Avenue East Cobb
Client: Cousins Properties Incorporated
Principal Consultants: Dougherty Schroeder & Associates, Inc. (design consultant to architect of record CMH, Inc.)
Post Properties, Inc., landscape architect
Ramon Luminance Design - Ramon Noya, lighting design
Hardin Construction, general contractor

Pinnacle Hills Promenade
Client: General Growth Properties, Inc.
Principal Consultants: Dougherty Schroeder & Associates, Inc. (full architectural services; architect of record)
Site Solutions, landscape architect
Bliss Fasman, Inc., lighting design
Huie Design, graphic design
Pacificom Multimedia, Inc., 3D graphic design
KLG Consolidated, mep
Shenberger & Associates, Inc., structural engineer
CEI Engineers. civil engineer
Spiker Baldwin, Inc., specifications

The Forum at Sunnyvale
Client: Forum Development Group
Principal Consultants: Dougherty Schroeder & Associates, Inc. (master planning & architectural design services)
Site Solutions, Inc., hardscape/landscape architect
BKF Engineers, civil engineer (retained by owner)
Standard Pacific, housing developer
MVE & Partners, Architect, housing design

Gulf Coast Town Center
Client: CBL & Associates Properties, Inc.
Principal Consultants: Dougherty Schroeder & Associates, Inc. (full architectural services; architect of record)
Site Solutions, landscape architect
Bliss Fasman, Inc., lighting design
Huie Design, graphic design
Brady & Anglin Engineers, mep
Shenberger & Associates, Inc., structural engineer

DUANY PLATER-ZYBERK & COMPANY

Tannin
Client: George Gounares
Principal Consultants: Duany Plater-Zyberk & Company, master planner
Caruncho, Martinez & Alvarez, residential and pool house architect
Khoury-Vogt Architects, town square designer
Michael Lykoudis, town center architect

Amelia Park
Client: Hometown Neighborhoods, Inc.
Principal Consultants: Duany Plater-Zyberk & Company, master planner
Oscar Machado, architectural and urban design consultant
Julie Sanford, town architect

I'On
Client: The I'On Company-Vince Graham, developer
Principal Consultants: Duany Plater-Zyberk & Company in collaboration with Dover-Kohl & Partners, master planner
Seamon Whiteside & Associates, civil engineer/ planner

Habersham
Client: Habersham Land Company- Robert Turner with Stephen Davis, developers
Principal Consultants: Duany Plater-Zyberk & Company, master planner
Steven Fuller Design Traditions, residential architect
Moser Design Group, residential architect
Historical Concepts, residential architect

Rosemary Beach
Client: The Rosemary Beach Land Company and Leucadia Financial Corporation
Principal Consultants: Duany Plater-Zyberk & Company, master planner
Keith LeBlanc Landscape Architecture, Inc., landscape architect
Aurora Civil Engineering, Inc., traffic engineer

Alys Beach
Client: Ebsco Gulf Coast Development
Principal Consultants: Duany Plater-Zyberk & Company, master planner
Khoury-Vogt, town architect
Glatting, Jackson, Kercher, Anglin, Lopez, Rinehart, traffic engineer
Moore Bass, civil engineering
Douglas Duany, landscape designer
James Wassel, renderer

ELKUS MANFREDI ARCHITECTS

35 and 40 Landsdowne Street
Client: Forest City Commercial Group, Millennium Pharmaceuticals Inc.
Principal Consultants: Elkus Manfredi Architects, master planning, architectural and interior design
AHSC-McLellan & Copenhagen, laboratory consultants
Halvorson Design Group, landscape designer
McNamara/Salvia, Inc., structural engineer
SEi Companies, mep engineer
Kaplan Architectural Lighting, lighting designer
Turner Construction Company, general contractor, 35 Landsdowne
Walsh Brothers Construction, general contractor, 40 Landsdowne

100 Cambridge Street/Bowdoin Place
Client: MassDevelopment/Saltonstall Building Redevelopment Corporation
Principal Consultants: Elkus Manfredi Architects, master planning, architectural and interior design
Sasaki Associates, associate architect
Brown, Richardson & Rowe, landscape designer
Daylor Consulting Group, Inc., civil engineer
LeMessurier Consultants, structural engineer
Vanderweil Engineers, Inc., mep engineer
Schweppe Lighting Design, Inc., lighting designer
Design, Inc., signage designer
Suffolk Construction Company, general contractor

ELS ARCHITECTURE AND URBAN DESIGN

California Theatre
Client: The Redevelopment Agency of the City of San Jose, Packard Humanities Institute
Principal Consultants: ELS Architecture and Urban Design, feasibility study, and architect
Weeks and Day, original (1927) architects
Rutherford & Chekene Consulting Engineers, structural
The Engineering Enterprise Consulting Engineers, electrical engineer
Guttmann & Blaevoet Consulting Engineers, mechanical
Brian Kangas Faulk, civil
Auerbach Pollock Friedlander, theatrical consultant
Auerbach Glasow, lighting consultant
Charles M. Salter Associates Inc., acoustical consultant
Davis Langdon Adamson, cost consultant
Hughes Associates, Inc., life safety consultant
TEECOM, telecom and security consultant
Top Flight, specifications consultant
Kate Keating Associates, Inc., signage consultant
James Goodman, decorative paint colors consultant
A.T. Heinsbergen with ELS Architects, interior furnishings and fabrics
Catalyst, landscape consultant
Rudolph & Sletten, construction manager
Swinerton Builders, general contractor

Church Street Plaza
Client: Arthur Hill & Co., LLC
Principal Consultants: ELS Architecture and Urban Design, master plan of district, design architect for Main Pavilion and parking structure
DeStefano + Partners, architect of record for Main Pavilion
Cooper Carry Architects, Hilton Hotel architects
OWP/P, 909 Davis (offices) architect
Walker Parking, parking structure architect
Nagel, Hartray, Danker, Kagan, McKay Architects, Borders Book Store architect
Optima, Inc., condominiums architect
Thorton-Tomasetti/Engineers, structural engineers
Environmental Systems Design, mep
Teska Associates, Inc, landscape
Bovis Construction Corp., general contractor

The Village at Merrick Park
Client: General Growth Properties, Inc.
Principal Consultants: ELS Architecture and Urban Design— concept master plan and architect for retail buildings and 2,000-car parking structure
SWA Group, landscape architect
Perkins & Will, office and residential buildings
Diedrich/NBA, Neiman Marcus store
Callison Architects, Nordstrom store
Sussman/Preja & Co., signage and graphics
Laurin B. Askew, Jr., of Monk LLC, client concept designer
The Lathrop Company, Turner Construction Company, contractor,

The City of Sunnyvale Downtown Design Plan
Client: City of Sunnyvale
Principal Consultants: ELS Architecture and Urban Design, urban design
Keyser Marston Associates, Inc., economist
Fehr & Peers Associates, Inc., transportation consultant
Architecture Models, Inc., model makers

FIELD PAOLI

Beverly Canon
Client: City of Beverly Hills, California
Principal Consultants: Field Paoli, Executive Architect
Jacques Verliden, Crate & Barrel architect
KPFF , structural engineer
Parkitects, parking consultants
Fong, Hart, Schneider Partners, landscape architects
W.E. O'Neil Construction Co., general contractor

On Broadway
Client: BHV Innisfree Ventures I, LLC
Principal Consultants: Field Paoli, architectural design
International Parking Design, parking structure
Freedman Tung & Bottomley, landscape architect
Architectural Lighting Design, lighting consultant
Fehlman Labarre, theater interiors

Victoria Gardens
Client: Forest City West Commercial Inc., Lewis Operating Co.
Principal Consultants: Field Paoli, master planning and concept design, building design
KA Inc., Architecture, executive architect
Altoon + Porter, executive design architect
Elkus Manfredi Architects, design architect
MDS Consulting, civil engineer
SWA, landscape design
Redmond Schwarz Mark, signage and graphics
City of Rancho Cucamonga Redevelopment Agency, city consultant

The Streets of Tanasbourne
Client: Continental Real Estate Companies
Principal Consultants: Field Paoli, conceptual design, architectural design
MESA Design, landscape
KPFF, structural engineer
Architectural Lighting Design, lighting

FXFOWLE ARCHITECTS, PC

The Helena Apartment Building
Client: The Durst Organization/Rose Associates
Principal Consultants: FXFOWLE ARCHITECTS, PC, architecture and sustainable Design
Flack + Kurtz, mep
Severud, Structural
Dagher Associates, living machine

The New York Times Building
Client: City Ratner Companies/The New York Times
Principal Consultants: FXFOWLE ARCHITECTS, PC, architecture
Renzo Piano Building Workshop, architecture
Flack + Kurtz, mep
Thorton Tomasetti, structural
Gensler, interiors
HM White Site Architects, landscape

Whitman School of Management
Client: Syracuse University
Principal Consultants: FXFOWLE ARCHITECTS, PC, architecture
Severud, structural
Flack + Kurtz, mep

Lincoln Center Redevelopment
Client: Lincoln Center for the Performing Arts
Principal Consultants: FXFOWLE ARCHITECTS, PC, architecture
Diller Scofidio + Renfro, architecture
Arup, mep

Tianjin Tower
Client: undisclosed
Principal Consultants: FXFOWLE ARCHITECTS, PC, architecture

Dosflota Multipurpose Complex Master Plan
Client: US CapitalInvest Bancorp
Principal Consultants: FXFOWLE ARCHITECTS, PC, architecture and urban design

Renaissance Place Redevelopment Plan
Client: Conroy Development Company
Principal Consultants: FXFOWLE ARCHITECTS, PC, architecture and urban design

GLATTING JACKSON KERCHER ANGLIN LOPEZ RINEHART, INC.

Hollis Garden
Client: City of Lakeland
Principal Consultants: Glatting Jackson Kercher Anglin Lopez Rinehart, Inc., park and public space planning, landscape architecture
Wallis Murphy Boyington Architects, Inc., architecture
Chastain-Skillman, Inc., engineer
Alex Piper, P.E., electrical engineer

Broad Street Park
Client: Baldwin Park Development Company
Principal Consultants: Glatting Jackson Kercher Anglin Lopez Rinehart, Inc., landscape architecture, urban design, environmental services, construction observation, relocation of trees
David G. Kittridge, PE, structural engineer
Prevost Irrigation Design, irrigation design
Hall Fountains, fountain mechanical

Park Avenue Streetscape
Client: City of Winter Park
Principal Consultants: Glatting Jackson Kercher Anglin Lopez Rinehart, Inc., planning, landscape architecture, transportation planning
Dover Kohl & Partners, urban design
GAI, civil engineer

The Heights
Client: Leslie Land Corporation
Principal Consultants: Glatting Jackson Kercher Anglin Lopez Rinehart, Inc., urban design, landscape architecture
Heidt and Associates, civil engineering
Fowler White Gillen Boggs Villareal & Banker, PA, legal counsel
Roberts Communication & Marketing, communications

GOODY CLANCY & ASSOCIATES

Fort Point Channel
Client: Boston Redevelopment Authority, Fort Point Channel Abutters Group
Principal Consultants: Goody Clancy & Associates, master planning
Vanasse Hangen Brustlin, Inc., watersheet permitting and quality
Vine Associates, marine infrastructure
FXM Associates, economic and market analysis
Byrne McKinney & Associates, implementation organization
Waterfront Center, charrette
Transportation Alternatives, water transportation

Assembly Square
Client: Assembly Square Limited Partnership (a collaboration of Gravestar, Inc. and Taurus New England Investments Corporation)
Principal Consultants: Goody Clancy & Associates, feasibility, architectural design, local permitting
Meredith & Grew, Inc., marketing
Atlantic Retail Properties, marketing
CLF Ventures Inc., environmental, transportation, community
Vanasse Hangen Brustlin, Inc., transportation, civil engineering
Carol R. Johnson Associates, Inc., landscape
Gibbs Planning Group, retail analysis
Connery Associates, fiscal impact
Reese Fayde & Associates, affordable housing
Woodard & Curran, environmental remediation
Gregory & Associates, legislative consultant
Holland & Knight LLP, legislative consultant
Nutter, McClennen & Fish LLP, legal counsel

North Allston Strategic Framework for Planning
Client: Boston Redevelopment Authority
Principal Consultants: Goody Clancy & Associates, planning
Volmer, transportation
Byrne McKinney Associates, real estate economics
Community Design Partnership, implementation strategies

LESSARD GROUP INC.,

Trump Plaza
Client: Cappelli Enterprises, Inc.
Principal Consultants: Lessard Group Inc., architect
HRH Construction, general contractor
Kellard Engineering, civil engineer
Tadjer-Cohen-Edelson Associates, Inc., structural engineer
SESI Consulting Engineers, P.C., geotechnical engineer
Edwards & Zuck, mechanical engineer
Cerami & Associates, acoustical
Lessard Commercial Inc., interiors
Desman Associates, parking
Lerch, Bates & Associates, elevators
Israel Berger & Associates, exterior envelope/fenestration/water proofing/roof

Springfield Town Center
Client: KSI Services, Inc.
Principal Consultants: Lessard Group Inc., architect
Sasaki & Associates, landscape

Canton Crossing
Client: KSI Services, Inc.
Principal Consultants: Lessard Group Inc., master planning
Whitney Bailey Cox & Magnani, LLC, civil engineer

National Harbor
Client: The Peterson Companies, K. Hovnanian Homes and McDaniels Homes
Principal Consultants: Lessard Group Inc., architect-in-charge, master planning, planning analysis and feasibility study, including building layout, density study and massing/height considerations
Tina Woods Smith, survey engineer
KTA Group, Inc., mechanical engineer
Tadjer-Cohen-Edelson Associates, Inc., structural engineer
Lawrence G. Perry, AIA, fair housing
Applied Fire Protection Engineering, code
Wyle Laboratories, acoustical engineer

LOONEY RICKS KISS ARCHITECTS

Jefferson at Providence Place
Client: JPI Development Partners, Inc.
Principal Consultants: Looney Ricks Kiss Architects, design architect, interior design, land planner
William F. Jervis, architect of record
Carol R. Johnson Associates, landscape architect
O. Ahlborg & Sons, Inc., builder

FedExForum
Client: Memphis Public Building Authority
Principal Consultants: Ellerbe Becket in association with Looney Ricks Kiss Architects, architecture
Looney Ricks Kiss Architects in collaboration with John F. Williams Architects, Inc., interior architecture
Looney Ricks Kiss Architects in collaboration with Bounds and Gillespie Architects, garage architecture
Looney Ricks Kiss Architects in collaboration with Self Tucker Architects, exterior skin architecture
Looney Ricks Kiss Architects, urban design, graphic design and theming
Ellerbe Becket, structural engineering, mechanical engineering, electrical engineering
Burr & Cole Consulting Engineers, structural engineering
ABS Consultants, structural engineering
Office of Griffith C. Burr, mechanical engineering, plumbing
Gala Engineering, mechanical engineering, plumbing
Shappley Design Consultants, fire protection
Liles Engineering Design Consultants, electrical engineering
Dunning-Martin Engineering, electrical engineering
PDR Engineers, civil engineering
Tetra Tech, civil engineering
Toles & Associates, infrastructure/civil engineering
Jackson Person & Associates, landscape architecture
Clark + Dixon Associates Architects, historic building survey
EnSafe, environmental consulting
Lerch Bates & Associates, Inc., vertical circulation
Walker Parking Consultants, garage consultant
M.A. Mortenson, general contractor

Ave Maria Town Center
Clients: Barron Collier Companies in partnership with Ave Maria University
Principal Consultants: Looney Ricks Kiss Architects, town core and Main Street planning; design of various town center mixed-use buildings
Wadsworth-O'Neal Engineering, Inc., mep and fp engineering
Liebl & Barrow Engineering, Inc., structural engineering
Bruce Howard Associates, landscape architecture
Wilson Miller, Inc., civil engineering, entitlements/site engineering
Williamson & Associates, waterproofing consultants
WJHW, acoustical consultant
Illuminating Concepts, exterior lighting designer

Ross Bridge Village Center
Client: Daniel Corporation
Principal Consultants: Looney Ricks Kiss Architects, master planning, architectural design
Holcombe, Norton & Pritchett, Inc., planning and landscape architecture
Walter Schoel Engineering Company, Inc., civil engineering
Lane Bishop York Delahay, Inc., structural engineering

Thornton Park
Client: name withheld
Principal Consultant: Looney Ricks Kiss Architects, architectural design
Alliance Structural Engineering, structural engineering
TLC-Engineering for Architecture, mep engineering
HDR, traffic engineering

GAI Consultants, civil engineering
Universal Engineering Services, geotechnical
Dix Lathrop & Associates, landscape architecture
Hardin Construction, contractor

MBH ARCHITECTS

West Hollywood Gateway
Client: J.H. Snyder
Principal Consultants: Principal Consultants: MBH Architects, Executive Architect, tenant architect for Best Buy and Target, LOD exhibits, leasing plans, design criteria book
ING Clarion, new owner
Jerde Partnership, design architect
Studio Mark, graphic design
Kaplan Partners, lighting design
EDAW, landscape architects
Swinerton Builders, general contractor

Marina University Villages
Client: Shea Properties
Principal Consultants: MBH Architects, architectural design, concept design, site planning, schematics, design development
Guzzardo Partnership, landscape
RBF, civil engineering

The Town Center at Levis Commons
Client: Dillin Development, Hill Partners
Principal Consultants: MBH Architects, design architect, design of master-planning, design of town center
Collaborative Group, executive architect
LKL Engineers Limited, structural
Mechanical Design Associates, Inc., mechanical / electrical
Fellar Finch, civil
Rudolph Libbe, cm
Mesa Design, landscaping
Kaplan, lighting design

200 Brannan
Client: Lennar Communities
Principal Consultants: MBH Architects, executive architect and design architect for this building, master planning of entire 4-building project with Kwan/Henmi Architects.
Luk & Associates, civil engineer
Nishkian Menninger, structural engineer
MPA Design, landscape architect
Charles M. Salter & Associates, acoustical consultant
CM & D, construction management
Bovis Lend Lease, contractor

MCLARAND VASQUEZ EMSIEK & PARTNERS

Fruitvale Village
Client: Fruitvale Development Corporation
Development Partners: BART, City of Oakland
Principal Consultants: McLarand Vasquez Emsiek & Partners, Inc., Urban Planning & Architecture
Fong, Hart Schneider Partners (prime landscape); Pattillo & Garrett Associates, landscape
Luk, environmental consultant
EQE Structural Engineers, structural engineering
Design Electric, electrical engineering
Hickey, plumbing
Bay City Mechanical, Inc., HVAC
Michael Willis & Associates, space planners
James E. Roberts-Obayashi Corporation, contractor

The Promenade at Rio Vista
Client: PLC Greystone Apartments
Principal Consultants: McLarand Vasquez Emsiek & Partners, Inc., Land planning & Architecture
Lifescapes International, Inc., landscape
DMC Engineering, civil
Group M Engineers, structural engineering
Helix Electric, Inc., electrical engineering
Parks Engineer, plumbing
LDI Heating & Air Conditioning, mechanical
Francis Krahe Associates, lighting lonsultan
Greystone Multi-Family Builders, Inc., contractor

Hollywood & Vine
Client: Legacy Partners and Gatehouse Capital
Principal Consultants: McLarand Vasquez Emsiek & Partners, Inc., Urban planning & Architecture
Rios Clementi Hale Studio, landscape
Fuscoe Engineering, civil engineering
DCI Engineers, structural engineering
HKA Pacific Parking Consultant, parking
Roschen Van Cleve Architects, historical & hollywood urban design consultant
Sussman Prejza, signage and graphics
Webcor Builders, contractor

Uptown Oakland Development
Client: Forest City Residential Development
Principal Consultants: McLarand Vasquez Emsiek & Partners, Inc., Urban Design & Architecture; Peter Calthorpe Associates, Urban Planners
Ken Kay Associates, master landscape planner
Korve, Inc., civil engineering
LSA Associates, Inc., EIR
KPFF Consulting Engineers, structural engineering
FARD Engineers, Inc., eechanical / electrical engineering
James E. Roberts-Obayashi Corporation, contractor

Douglas Park
Client: Boeing Realty Corporation
Principal Consultant: McLarand Vasquez Emsiek & Partners, Inc., Urban planning & Architecture

Tralee
Client: Bancor Properties LLC
Principal Consultants: McLarand Vasquez Emsiek & Partners, urban design, planning, architecture
The Guzzardo Partnership, Inc., landscape
Carlson, Barbee & Gibson, Inc., civil engineer

PAPPAGEORGE/HAYMES LTD.

The Glen Town Center
Client: Oliver McMillan, Kimball Hill Homes
Principal Consultants: Pappageorge/Haymes Ltd., planning, architectural design
Douglas Hoerr Landscape Architecture, landscape
Samartano & Company, structural
Cosentini Associates, mep
Cowhey Gudmundson Leder, civil

600 North Lake Shore Drive
Client: Belgravia Group Ltd. & SandZ Development Company, Inc.
Principal Consultants: Pappageorge/Haymes, Ltd., architectural design
Samartano & Company, structural
Khatib & Associates, mep
Eriksson Engineering Associates, Inc., civil
Hitchcock Design Group, landscape
Wolf Clements & Associates, landscape

Block X
Client: The Thrush Companies
Principal Consultants: Pappageorge/Haymes Ltd., architectural design
Peter R. Krallitsch & Associates, structural
McClier Corporation, civil
Joe Karr & Associates, landscape

Museum Park
Client: A joint venture of The Enterprise Companies & Central Station Development Corporation
Principal Consultants: Pappageorge/Haymes Ltd., architectural design
Samartano & Company, structural
Stearn Joglekar & Associates, structural
Rosenblatt Associates, mep
Cosentini Associates, mep
Eriksson Engineering Associates, Inc., civil
Daniel Weinbach & Associates, landscape
Joe Karr & Associates, landscape
Horvath & Reich, exterior wall
Jenkins & Huntington, elevator
Charter Sills & Associates, lighting

Kinzie Park
Client: The Enterprise Companies, The Habitat Company
Principal Consultants: Pappageorge/Haymes Ltd., architectural design
Peter R. Krallitsch & Associates, structural, townhouses
Chris P. Stefanos Associates, structural, midrise
GKC/EME, LLC, mep
Eriksson Engineering Associates, Inc., civil
Joe Karr & Associates, landscape

PERKOWITZ + RUTH ARCHITECTS

Bridgeport Village
Client: Center Oak Properties, LLC
Principal Consultants: Perkowitz + Ruth Architects, schematic design, design development
Candera, lighting
Berger Partnership, landscape
Opus A & E, engineer

Mercantile West
Client: DMB Ladera, Westar Associates
Principal Consultants: Perkowitz + Ruth Architects, architectural design and construction services
Land Concern, landscape
Lighting Design Alliance, lighting
ANF & Associates, engineer

Buena Park Downtown
Client: Festival Company, Pritzker Realty, Krikorian Premiere Theatres
Principal Consultants: Perkowitz + Ruth Architects, design through construction phase, including documents and administration
EDAW, Inc., landscape architect
ANF & Associates, structural engineer

RETZSCH LANAO CAYCEDO ARCHITECTS

Royal Palm Office Building
Client: RLC Development, LLC
Principal Consultants: Retzsch Lanao Caycedo Architects, design through construction administration and construction management, interior design
Donnell Duquesne Albaisa, P.A., structural engineer
Thompson Engineering Consultants, mep engineer
EcoPlan, landscape architect
Icon Design Group, contractor

Cypress Park West, Phase II
Client: TIAA-CREF (Savannah Teachers Properties), CB Richard Ellis as agent for teachers
Principal Consultants: Retzsch Lanao Caycedo Architects, site planning, architectural design through construction administration for new building and parking structure
Johnson Structural Group, structural engineer
Thompson Engineering Consultants, mep engineer
Sun Tech Engineering, civil engineer
Red Eye Design, landscape architect
Phillips Fire, fire protection
Itasca Construction Associates, contractor

Fifth Avenue Place, Phase II
Client: Mocal Enterprises
Principal Consultants: Retzsch Lanao Caycedo Architects, site feasibility and schematic design through construction administration
O'Donnell, Naccarato, Mignogna & Jackson, Inc., structural engineering
Thompson Youngross Engineering Consultants, mep engineer
Caufield & Wheeler, Inc., civil engineer
A. Grant Thornbrough & Associates, landscape architect
Phillips Fire, fire protection
TBD, contractor

The Pointe at Middle River
Client: Brenner Real Estate Group
Principal Consultants: Retzsch Lanao Caycedo Architects, site planning and architectural design, design development, construction documents and construction administration
Donnell Duquesne Albaisa, P.A, structural engineering
Thompson Youngross Engineering Consultants, mep engineer
Sun Tech Engineering, civil engineer
EcoPlan, landscape architect
Phillips Fire, fire protection
TBD, contractor

RTKL

LaQua Tokyo Dome City
Client: Takenaka Komuten KK
Principal Consultants: RTKL Associates Inc., architecture
Takenaka, construction/structural engineering/landscape architecture
Takenaka and Kandeko, electrical engineering
Sangi Kogyo, TAK Evac Johnson Controls, mechanical/plumbing
Matsushita, lighting systems

Principe Pio
Client: Riofisa
Principal Consultants: RTKL Associates Inc., architecture
Estudio Fernandez del Amo, architect of record
IDOM S.A., mep engineering, structural engineering
Eralan, main contractor
Mero, Germany, glass subcontractor
Michael Schlaich, glass engineer

Downtown Brea Redevelopment District
Client: City of Brea and CIM Group LLC
Principal Consultants: RTKL Associates Inc., master planning, urban design, tenant design criteria, architectural guidelines, environmental graphic design, design review, signage presentation
Robin Faulk, events planning
Lauren Melendrez Associates and AHBE, construction documents
Olson Company, townhouse design

Zha Bei/The Hub International Lifestyle Centre
Client: Forrester Group
Principal Consultants: RTKL Associates Inc., architecture, environmental graphic design, master planning
SIADR (Shanghai Institute of Architectural Design and Research, local architect
WSP Hong Kong, Ltd. and SIADR, mep engineering, structural engineering
Chroma33, architectural lighting design

SASAKI ASSOCIATES, INC.

Addison Circle Park
Client: Town of Addison
Principal Consultants: Sasaki Associates, Inc., planning, urban design, landscape architecture, civil engineering
Jim Duffy, construction manager
Georgia Fountain Company, Inc., fountain design
Irritech Corporation, irrigation design
Campos Engineering, Inc., mep engineers
Gary Cunningham, Cunningham Architects, pavilion design

Thu Thiem New Urban Center
Client: Investment and Construction Authority for Thu Thiem
Principal Consultants: Sasaki Associates, Inc., planning, urban design, landscape architecture

Charleston Waterfront Park
Client: City of Charleston
Principal Consultants: Sasaki Associates, Inc., master planning, urban design, landscape architecture, civil/marine engineering
Holladay, Coleman and Associates, electrical engineers
David Carsen, structural engineer
LAW Engineering and Environmental Services, Inc. and MACTEC Engineering and Consulting, geotechnical engineers (In 2002, LAW merged with MACTEC.)
Edward Pinckney Associates, Inc., local landscape architects
CMS Collaborative, Inc., fountain mechanical/electrical design and engineering
Ruscon Construction Company, Inc., contractor

Detroit Riverfront Civic Center Promenade
Client: City of Detroit Parks Department
Principal Consultants: Sasaki Associates, Inc., planning, landscape architecture, civil engineering
The Albert Kahn Collaborative, client/architect
NTH Consultants, Ltd., environmental engineer
Tucker, Young, Jackson, Tull, Inc., civil engineer
Snell Environmental Group (SEG), surveyor

SEH (SHORT ELLIOTT HENDRICKSON INC.)

Mound Public Safety Facility
Client: City of Mound, Minnesota: Fire Station and Police Department
Principal Consultants: SEH (Short Elliott Hendrickson Inc.), feasibility study, community open house, architectural, landscape, site civil, mechanical, electrical
Bob Perzel, artist

Heart of Anoka Commuter Village Master Plan
Client: City of Anoka, Minnesota
Principal Consultants: SEH (Short Elliott Hendrickson Inc.), site inventory and analysis, master planning, transportation planning, civil engineering, public facilitation, landscape architecture
BRT Architects, architecture

Maxfield Research, Inc., real estate economics, market research

Loring Bikeway and Park
Client: City of Minneapolis, Minnesota
Principal Consultants: SEH (Short Elliott Hendrickson, Inc.), site analysis, preliminary and final bridge, trail and park design, public art consulting, public advocacy
Lisa Elias, decorative railing for Pocket Park

I-35W Access Project
Client: Hennepin County
Principal Consultants: SEH (Short Elliott Hendrickson Inc.), public facilitation, transportation planning, urban design, landscape architecture
HDR, engineering
Milo Thompson, architecture,
Gary Hallman, artist

Gateway Centre
Client: Gateway Centre, LLP
Principal Consultants: SEH (Short Elliott Hendrickson, Inc.), architectural design
Nicol Associates, Inc., structural
Innovative Mechanical Systems, Inc., mechanical/plumbing
Architectural Engineering Design Group, electrical

SWA GROUP

Lite-On Electronic Headquarters
Client: Artech Inc.
Principal Consultants: SWA Group, full landscape architectural design services for hardscape, softscape, and water elements
Innerscape design, local landscape architect for construction document production
Artech, Inc., building architect

PPG Place
Client: Hillman Properties, Inc.
Principal Consultants: SWA Group, urban design, landscape architecture
R.M. Gensert, structural engineer
WET Design, fountain consultant
IKM, Inc., architect/designer for renovation

Hangzhou HuBin Commerce & Tourism District Redevelopment Master Plan
Client: Hangzhou HuBin Commerce & Tourism District
Principal Consultants: SWA Group, urban redevelopment, master planning, Phase 1 design development, construction phase services
Jerde Partnership International, architecture, consultation on planning
ZSADI (Zhejing Southern Architectural Design Institute), architecture

Lewis Avenue Corridor
Client: City of Las Vegas
Principal Consultants: SWA Group, urban planning, site planning design, construction phase services (full landscape architectural services)
Russ Mitchell and Associates, irrigation
Poggemeyer Design Group, civil and structural
JBA Consulting Engineers, electrical
Fountain People, fountains
Comprehensive Planning Division, City of Las Vegas, planning

Santana Row
Client: Federal Realty Investment Trust (private)
Principal Consultants: SWA Group, full landscape architectural design services, including water features and construction administration

SB Architects (Sandy & Babcock), building architect
BAR Architects (Backen Arrigoni & Ross, Inc.), building architect
Steinberg Architects (The Steinberg Group), building architect
April Philips Design Works, additional landscape architecture design

SWABACK PARTNERS, PLLC

DC Ranch
Client: DMB
Principal Consultants: Swaback Partners, pllc, overall master plan, architectural guidelines, and special community features,
Various Structures by Multiple Architects

Biosphere 2
Client: Decisions Investments Corporation, CB Richard Ellis
Principal Consultant: Swaback Partners, pllc, conceptual planning and additional architectural facility alternatives

Scottdale Hangar One
Client: Scottsdale Hangar One
Principal Consultants: Swaback Partners, pllc, architects
Tihany International, interior design
Studio V, interiors technical support
Paul Koehler Engineers, structural
Lockwood Greene, mep
Automation IQ, automation technology
Steve Martino & Associates, landscape architecture

The Village of Kohler
Client: Kohler Co.
Principal Consultants: Swaback Partners, pllc, overall master planning, detailed site planning, architects for major buildings, special community features
Various Structures by Multiple Architects

Las Palomas
Client: Abigail Properties
Principal Consultants: Swaback Partners, pllc, site planning and architecture
Studio V, interior design
Forest Richardson, golf course architect

Marana Master Plan
Client: Town of Marana
Principal Consultant: Swaback Partners, pllc, landplanning, urban design, site design, design guidelines, architectural visioning, public process-participation, zoning documentation preparation

THOMAS BALSLEY ASSOCIATES

J-City
Client: Mitsui Fudosan Co. Ltd.
Principal Consultants: Thomas Balsley Associates, full landscape design services through documentation and construction over slab
Yamashita Sekkei, architect

Capitol Plaza
Client: Witkoff Group/Adell Corporation
Principal Consultants: Thomas Balsley Associates, full landscape design services through documentation and construction
Costas Kondylis & Associates, architect
Rosenwasser/Grossman, mep engineer

Pacific Design Center
Client: Cohen Brothers Realty Corp.
Principal Consultants: Thomas Balsley Associates, full landscape architectural design services through documentation and construction
Melendrez Design Associates, fandscape architectural construction documentation and construction services
Land Design Consultants, engineering services
Selbert Perkins Design, Graphic Consultant
Kaplan Partners Architectural Lighting, lighting designer
Crain & Associates, traffic engineers
Fluidity Design Consultants, Inc., water features

Riverside Park South
Client: Riverside South Planning Corp.
Principal Consultants: Thomas Balsley Associates, urban design/landscape architect
Skidmore, Owings & Merrill, urban planner
HNTB, engineering services
Olko Engineering, engineering services
Ysrael Senuk, P.C., engineering services
Philip Habib Associates, engineering services
Han-Padron Associates, marine engineering services
Parsons Brinkerhoff, construction manager
Lehrer McGovern Bovis, cost estimator/construction manager

World Trade Center Plaza
Client: World Trade Center Associates/Nikken Sekkei
Principal Consultants: Thomas Balsley Associates, full landscape design services through documentation and construction over slab
Nikken Sekkei, architects
Mancini Duffy, architects
LPA Inc., lighting

THOMAS P. COX: ARCHITECTS, INC.

Stapleton Town Green
Client: Forest City Properties
Principal Consultants: Thomas P. Cox: Architects, Inc., architect

North Crescent
Client: Playa Vista, Fairfield Residential
Principal Consultants: Thomas P. Cox: Architects, Inc., architecture and planning
ima+design, landscape architect
Style Interiors, interior design

Grand Avenue Competition
Client: Forest City Enterprises
Principal Consultants: Thomas P. Cox: Architects, Inc., architect
A.C. Martin, architect
Calthorpe Associates, planning
Civitas, landscape, landscape architect

Westgate-Pasadena
Client: Sares Regis
Principal Consultants: Thomas P. Cox: Architects, Inc., architect
Melendrez, landscape architect

Index by Project

100 Cambridge Street/Bowdoin Place, Boston, MA, **150**
200 Brannan, San Francisco, CA, **216**
35 and 40 Landsdowne Street, University Park at MIT, Cambridge, MA, **146**
4025 Connecticut Avenue, Park Hill Condominiums, Washington, DC, **111**
5225 Maple Avenue, Dallas, TX, **44**
600 North Lake Shore Drive, Chicago, Illinois, **228**

Addison Circle Park, Addison, TX, **258**
Allegro Tower Apartments, San Diego, CA, **34**
ALTA, The, Thomas Circle, Washington, DC, **109**
Alys Beach, Panama City, FL, **144**
Amelia Park, Fernandina Beach, FL, **139**
Aqua Condominiums, Panama City Beach, FL, **84**
Assembly Square, Somerville, MA, **190**
Ave Maria Town Center, Collier County, FL, **206**
Avenue East Cobb, The, Atlanta, GA, **133**
Ayala Center Greenbelt, Makati City, Philippines, **50**

Beth Israel Deaconess Medical Center & Master Plan, Boston, MA, **74**
Beverly Canon Mixed-Use Retail, Beverly Hills, CA, **162**
Biosphere 2, Oracle, AZ, **284**
Black Diamond, Pittsburg, CA, **120**
Block X, 1145 Washington, Chicago, IL, **229**
Botanica on the Green, Stapleton, CO, **298**
Bridgeport Village, Tualatin, OR, **234**
Brighton Village, Coolidge, AZ, **32**
Broad Street Park, Baldwin Park, FL, **180**
Buena Park Downtown, Buena Park, CA, **239**
Buena Vista, Bakersfield, CA, **60**

California Theatre, San Jose, CA, **154**
Canton Crossing, Baltimore, MD, **198**
Capitol Plaza, New York, NY, **292**
Caton's Walk, Georgetown, Washington, DC, **108**
Charleston Waterfront Park, Charleston, SC, **262**
Cheval Apartments on Old Katy Road, Houston, TX, **88**
Church Street Plaza, Evanston, IL, **156**
Citrus Plaza, Redlands, CA, **70**
City Hall Plaza Community Arcade and Government Center Master Plan, Boston, MA, **76**
City of Sunnyvale Downtown Design Plan, The, Sunnyvale, CA, **160**
Columbus Center, Boston, MA, **90**
Coyote Valley, San Jose, CA, **114**
Crescent Creek, Raytown, MO, **14**
Crescent Park Apartment Homes, Playa Vista, CA, **300**
Cypress Park West, Phase II, Fort Lauderdale, FL, **244**

Davis Building, The, Dallas, TX, **47**
DC Ranch, Scottsdale, AZ, **282**

Destin Commons, Destin, FL, **130**
Detroit Riverfront Civic Center Promenade, Detroit, MI, **264**
Dosflota Multipurpose Complex Master Plan, Moscow, Russia, **175**
Douglas Park, Long Beach, CA, **223**
Downtown Brea Development District, Brea, CA, **252**

Easton, Dallas, TX, **46**
Echelon I, Las Vegas, NV, **237**
Egyptian Lofts, San Diego, CA, **36**

Fashion Show, Las Vegas, NV, **22**
FedExForum, Memphis, TN, **204**
Fifth Avenue Place, Phase II, Boca Raton, FL, **246**
Figueroa Central Downtown, Los Angeles, CA, **236**
Flats at Rosemary Beach, The, Rosemary Beach, FL, **87**
Fort Point Channel, Boston, MA, **186**
Fort Washington Way Highway Reconfiguration, Cincinnati, OH, **80**
Fort Worth Master Plan, Fort Worth, TX, **122**
Forum at Sunnyvale, The, Sunnyvale, CA, **134**
Fruitvale Village, Oakland, CA, **218**
Fortnightly Neighborhood Master Plan, New Senior Community Center, Herendon, VA, **112**

Gateway Centre, Longmont, CO, **272**
Glen Town Center, The, Glenview, IL, **226**
Grand Avenue Competition, Los Angeles, CA, **302**
Grand Gateway, Shanghai, China, **52**
Grande Lakes Resorts, Orlando, FL, **58**
Grand Tier, The, 1930 Broadway, New York, NY, **104**
Gulf Coast Town Center, Fort Myers, FL, **136**

Habersham, Beaufort, SC, **141**
Hangzhou HuBin Commerce & Tourism District Redevelopment Master Plan, Hangzhou, China, **278**
Heart of Anoka Commuter Rail Village Master Plan, Anoka, MN, **268**
Helena Apartment Building, The, New York, NY, **170**
Heritage, The, New York, NY, **100**
Highlands of Lombard, The, Lombard, IL, **42**
Hollis Garden, Lakeland, FL, **178**
Hollywood & Vine, Hollywood, CA, **221**
Huntfield, Charles Town, WV, **107**

I'On, Mount Pleasant, SC, **140**
I-35WS Access Project, Hennepin County, **271**

J-City, Tokyo, Japan, **290**
Jefferson at Providence Place, Providence, RI, **202**

Kinzie Park, Chicago, IL, **232**
Knox Shopping Centre, Melbourne, Australia, **18**

Lakes at Thousand Oaks, The, Thousand Oaks, CA, **240**
LaQua Tokyo Dome City, Tokyo, Japan, **250**
Las Palomas, Puerto Penasco, Mexico, **287**
Lewis Avenue Corridor, Las Vegas, NV, **279**
Lincoln Center Redevelopment, New York, NY, **173**
Lite-On Electronic Headquarters, Taipei, Taiwan, **274**
Loring Bikeway and Park, City of Minneapolis, MN, **270**
Luxe Hills International Golf Community, Cheng Du, Sichuan Province, China, **116**

Marana Master Plan, Marana, AZ, **288**
Marina University Villages, Marina (Fort Ord), CA, **212**
Mather Building, The, Washington, DC, **106**
Mercantile West, Ladera Ranch, CA, **238**
Metropolitan Tower, Seattle, WA, **56**
Milan Lofts, Pasadena, CA, **237**
Morton Square, 600 Washington St., New York, NY, **102**
Motorola Global Software Group, Krakow, Poland, **26**
Mound Public Safety Facility, Mound, MN, **266**
Museum Park, Chicago, IL, **230**

National Cathedral School, Washington, DC, **110**
National harbor, National harbor, MD, **200**
New Longview, Lee's Summit, MO, **12**
New Town Theater District, St. Charles, MO, **10**
New York Times Building, The, New York, NY, **171**
North Allston Strategic Framework for Planning, Boston, MA, **192**
North Hills, Raleigh, NC, **66**
North Point, Cambridge, Boston, Somerville, MA, **91**

On Broadway, Downtown Redwood City, CA, **164**
Ottawa University Master Plan, Ottawa, KS, **15**

Pacific Design Center, West Hollywood, CA, **294**
Park Avenue Streetscape, Winter Park, FL, **182**
Park Laurel on the Prado, San Diego, CA, **38**
Parker Square, Flower Mound, TX, **128**
Pinnacle Hills Promenade, Rogers, AK, **132**
Pinnacle Museum Tower, The, San Diego, CA, **40**
Planning Design of Haikou West Coast, Haikou City, China, **28**
Pointe at Middle River, The, Oakland Park, FL, **248**
Power Plant Live!, Baltimore, MD, **69**
PPG Place, Pittsburgh, PA, **276**
Principe Pio, Madrid, Spain, **254**
Promenade at Rio Vista, The, San Diego, CA, **220**
Promenade Town Center, Pasco County, FL, **64**
Prudential Center Redevelopment, The, Boston, MA, **94**

Rarity Pointe Lodge and Spa, Knoxville, TN, **86**
Renaissance Place Redevelopment Plan, Naugatuck, CT, **176**
Residences at Kendall Square, The, Cambridge, MA, **92**
Residential Projects, **16**

Riverside Park South, New York, NY, **295**
Rollins Square, Boston, MA, **93**
Rosemary Beach, Panama City, FL, **142**
Ross Bridge Village Center, Birmingham, AL, **207**
Royal Palm Office Building, Boca Raton, FL, **242**

Santana Row, San Jose, CA, **280**
Scottsdale Hangar One, Scottsdale, AZ, **285**
Serta International Center, Hoffman Estates, IL, **30**
Shoppes at Blackstone Valley, The, Millbury, MA, **71**
Smart Corner, San Diego, CA, **39**
Solivita, Poinciana, FL, **62**
Southlake Town Square, Southlake, TX, **126**
Springfield Town Center, Springfield, VA, **196**
Streets of Tanasbourne, The, Hillsboro, OR, **168**
Suwon Gateway Plaza, Suwon, Korea, **54**

Tannin, Orange Beach, AL, **138**
The Heights, Tampa, FL, **184**
Thornton Park, Orlando, FL, **208**
Three Rivers Park, Pittsburgh, PA, **78**
Thu Thiem New Urban Center, Ho Chi Minh City, Vietnam, **260**
Tianjin Tower, Tianjin, China, **174**
Town Center at Levis Commons, The, Perrysburg (Toledo), OH, **214**
Tralee, Dublin, CA, **224**
Trump Plaza, New Rochelle, NY, **194**
Trump World Tower, 845 United Nations Plaza, New York, NY, **98**

Union Station, Union, NJ, **48**
University Villages, Marina, CA, **118**
Uptown Maitland West, Maitland, FL, **85**
Uptown Oakland Development, Oakland, CA, **222**

Victoria Gardens, Rancho Cucamonga, CA, **20**
Victoria Gardens, Rancho Cucamonga, CA, **166**
Village of Kohler, The, Kohler, WI, **286**
Village of Merrick Park, The, Coral Gables, FL, **158**

Walk, The Atlantic City, NJ, **68**
Walkers Brook Crossing, Jordan's Furniture, Home Depot, IMAX, North Reading, MA, **72**
West Hollywood Gateway, West Hollywood, CA, **210**
West Village, Dallas, TX, **127**
Westgate – Pasadena, Pasadena, CA, **304**
Whitman School of Management, Syracuse University, Syracuse, NY, **172**
Williams Walk at Bartram Park, Jacksonville, FL, **82**
World Trade Center Plaza, Osaka, Japan, **296**

Zha Bei/The Hub International Lifestyle Centre, Shanghai, China, **256**

Acknowledgments

The first edition of *Urban Spaces* was launched just about ten years ago. Over this decade we have had the privilege of showcasing the very best in urban design and development. This continuing series represents one of the most extensive pictorial collections of contemporary urban projects.

Our fourth edition of *Urban Spaces* is the result of the combined efforts and the cooperation of a talented group of professionals representing the best in their field.

The Urban Land Institute has again cosponsored the publication of *Urban Spaces*. The support and guidance of Rick Rosan, Rachelle Levitt, Gayle Berens, Karrie Underwood and Lori Hatcher were essential to publishing this series. Many thanks to this great organization for its continued interest in *Urban Spaces*.

John Dixon, once again, shepherded the editorial process. His descriptive prose and technical knowledge enhanced the presentation of the over 190 projects covered in the new edition. His skill in developing the essential elements of each project and interpreting the goals of the design firms expedited the exchange of copy and graphics and required a minimum of corrections and rewriting.

The imaginative and effective layouts of the images were produced by our expert designer and photo editor, Harish Patel, who was the point man in turning proofs around quickly and interpreting the proof marks and comments of the participating firms. He was ably assisted by our master of proofreading, John Hogan. Although we utilize all the state-of-the-art electronic communications and graphic systems, no software can substitute for the artistic ability required to create good visual design.

The production professionals at our printers, Gracie Xie and Avan Lee, are also to be thanked for their contribution to the production of this beautiful book. Their watchful eye and attention to detail through all the myriad printing processes ensured the timely completion of the book from disk through distribution.

Of course, there would not have been a fourth edition of *Urban Spaces* if it were not for the cooperation and interest of the architects, marketing executives and graphic directors representing the 37 firms featured in the new volume. We very much enjoyed working with them and became good phone friends and e-mail buddies.

Thanks again to all of you that made this possible.

Henry Burr
Publisher